Nature's Essence as Living Architect And An Alchemist Of Life

NatalDNA Diet and Precision of Allocation for Higher Intelligence and Optimal Health.

Natalie Manroe

DEDICATION

"In the pages of this book, we embark on a journey where Nature's Divine Essence gracefully weaves into the tapestry of human engineering as a cellular inseparable extension of Nature and all Living Matter and its power of infinite potential as the cultivator of Life Force Energy and creator of own magic, health, longevity, environment, life and last. Here, it steps into its true power, unveiling the source of creation, the unquestionable laws that govern existence, and the origin of its Mighty Force. With an unwavering claim to ownership over all Living Matter, this narrative unfolds the profound narrative of Nature as not only a Designer but also as an Architect and an Alchemist of Life. Through these words, we celebrate the architectural design created and bound by the Natural Laws for all Living Matter, yet boundless inspiration drawn from the cosmic symphony of creation and the price we pay for ignoring preordination and Divine Laws."

ACKNOWLEDGMENTS

With immense gratitude, I want to express my heartfelt thanks to my family, friends, every book, technical paper, story or journal; every person on my path that I treated as contributor or teacher, every reader who gave their time, Creative Force and all the Nature's given guidance and years of endless serendipities that taken me through the experience I had never imagined guiding me to write the transcript of Nature's Essence, Architectural Laws and connecting the thread within this book. Your unwavering support has been my greatest inspiration and the driving force behind every word penned. I couldn't have embarked on this journey without your love and encouragement. Here's to the incredible tapestry of connections among all life forms and all those wonderful souls that cross path my life.

CONTENTS

SYNOPSIS

Author breaks silence after overcoming many formidable health hurdles, observing struggles of others in many walks of life globally, investigating by asking questions and diving into the deepest secrets behind human design, Nature's destructive and transformative power breaking apart unlike ever before. Exploring all from the perspective of Nature as an Architect and Alchemist of life. Authors Unique design of each human body, its true potential, bio-availability, and true role as building blocks of life. Discover the essence of life's unspoken mechanics as we explore nature's role as living architect, creator and alchemist of life, which is the manual to our health, the health of our children, all species, the planet and the next stage in human evolution.

As we dive in and unveil undebatable laws of your unique design, answers to simple why, what, and how of nature's creation, offer unique insights into human engineering, precision narratives, stepping stones and the importance of Nature's allocated NatalDNA Diet, role, order, energy and purpose.

Learn the timeless wisdom of ancestral virtues, embrace a healthy and flexible mind-set, and connect with ancient knowledge. Know and understand how diet affects everything from minuscular to grand or from cellular to global. Find empowerment in understanding the Natural Laws, preordained health, Life Force energy, how nature made you and the simplicity of hidden power within you from cellular to collective, personal or global.

We look into modern science, ancient knowledge of nature's medicinal source and the power of immunity, plants, earth pharmacy, arrangements, order and herbs. We explore nature's unstoppable meddling, solutions beyond gut health, weight loss, skin, ageing and every health condition we inherit, the moods, the highs and lows, anxiety or joy, longevity, fertility, health and last. Do we become what we build or what we accumulate? Is all knowledge is wisdom? A comprehensive guide to priceless answers and solutions based on the nature of your unique design, in becoming an expert and take control of your health.

Natalie Manroe

1 CHAPTER

NATURE'S ENGINEERING AND THE TAPESTRY OF LIFE

Ever Living Mastery of Creation and its preordained power guided by Natural Laws thought of all that exists in Nature from minuscular to grand, way before any of us come to this world.

Just like Earth, the human body is designed by nature and bound by the same laws. And just like soil becomes what we put in it and has needs for its requirements according to what it will grow or the environment it will build, the body does too. The question is, what foods are compatible with unique to each nature-allocated genetic template to become the best cellular intelligence it can be? It's up to you, to make it your business to find out and up to you to neglect into decline or improve, prevent and maintain.

Ever Living Mastery of Creation and its preordained power guided by Natural Laws. Life Force fuels and thought of all that exists in nature from minuscular to grand. Just like the Earth, the human body is designed by The Force of Nature and bound by the same laws. And just like soil becomes what we put in it and has needs for its requirements according to what it will grow or the environment it will build, the body does too. The question is, what are the NatalDNA compatible foods? Why such foods must be precise and unique to each nature-allocated genetic NatalDNA template to become the best cellular intelligence it can be? Why are Natural Laws and Ever Living Mastery matters? Next level evolution of humanity. We dive into all details in this book. It's up to you to make it your business to find out and up to you to neglect into decline or improve, prevent and maintain.

The tapestry of Life can be described in many forms. But here, we explore Nature as a creator, architect, alchemist, scientist, manufacturer, project manager, software and hardware engineer, Cellular Intelligence, and an ongoing developer of its own design. Cellular Intelligence is equivalent to what humans developed and known as Artificial Intelligence. All of these geniuses at once. And how does your cellular health evolve into possibilities of Nature's architectural draft over time, displaying transcendent purpose and unity among all life? How does it all work and depend on each other? Nature holds the key to the best life supporting system model for all, the path for health, longevity, and abilities to manifest and harvest evolutionary advancements for all life forms. Just as an architect carefully designs and allocates resources for a structure, or a car designer considers the fundamental elements, including the fuel source, nature works similarly. It adheres to all-natural laws, including the vital principle of Order, to accommodate and harmonise all life. Removing what does not serve, corrupts, or interfere with its preordain purpose and design will give us solutions and answers we are looking for.

Without scrutinising each particle that enters our mouth, goes on our skin, or enters through the breath we take, we can never achieve equilibrium of life and health for ourselves and others. Such not without an in-depth understanding of nature's mechanics. Just like a car mechanic has to understand the fundamental building blocks of a car. Each must understand the fundamental building blocks of its nature given body with its unique built system. This is as important as the type, size, and precision of each screw that goes into Elon Musk's starship. The difference is that it is amassing so-called "wrong screws" and "misfits" manifest and display human health results that are visible now. Many pay little attention to it until health issues and illnesses that mostly appear in older age begin to manifest in younger people.

The answers we seek and the example of the transcendent model of the world we all want were always there. Its Globality of understanding was missing. This is because the evolution of the mind had to take place way before higher levels of understanding, advancing levels of cellular consciousness, powerhouse, and higher levels of spiritual intelligence becomes achievable. Just like our own growth, understanding is different at 20yr then it is at 40yr onwards. In nature's eyes, each century is like a school year, guiding us step by step. Nature designed us in such ways, so we grow with nature's pace and rhythm. However, since human nature and old habits in some are still overpowering, causes contradiction and clash, pushing us to grow in our views and understanding. Now that we can relate to technological advancements, True Laws and what goes into each tech, or manufacturing process, we can relate more to the particularity of ancient teachings and scripts having the possibility to grasp and relate to system mechanics and the Nature of Alchemy and Architectural design that goes into each living matter and defines the purpose behind Natural Laws and evolution of life.

Just as our understanding of technologies has grown, so has our understanding of technicality and precision that comes with it. For instance, as dumb as it sounds, but we know now about importance and the need for devices to be plugged into the electric powerhouse to charge, and the importance of the matching cable; or the obvious, such as precision and quality of material we choose for anything that we manufacture or build. Why do we expect the very nature of ourselves and dietary requirements for precision to be any different? We are about to unveil comprehending the depth behind the precision, Natural Laws and all the rules, order and why's that human body bound by and comes with. Part of evolutionary development and challenge in some way is to achieve emotional intelligence, comprehension and unity among species without force. Keep reminding ourselves that nature operates in real time, its own pace, nothing in nature is done by force and each human body is bound by the same laws. More so, when we apply force without understanding that force goes both ways and has tendency to push back. The challenge is to achieve evolutionary comprehension on the same level willingly, just as each cell in our body does: each different yet the same, unique, yet interdependent and matters.

Many analogies can lead us to the depth of enlightenment fuelled by spiritual and cellular intelligence as an inner core of human experience as a species in the form of the creation powerhouse of living matter. So, we can discover that, just as a mobile phone is useless without charge nor without a matchable charger or energy supply; architectural and alchemic pathways of our cellular abilities including its powerhouse require precision of such match too for us to see what is truly possible. Should we want to experience Nature's true magnificence which is hidden from within?

Each cell in your body is as unique as you in this life. The question we have to ask is what makes a population of complex living matter inside your body with a population higher than people on this planet work and co-exist so perfectly and what is not? Almost 40 trillion cells. This is 4.5 times more than the people on planet Earth. Cellular intelligence of such calibre possesses the power of knowing precisely what to do, functioning and co-existing with each other, its pre-ordained body system and organ. Nevertheless, we, who call ourselves an intelligent species, cannot manage our own system without a peaceful resolution and life-supporting environment for all life forms. Sadly, we take for granted, mistreat, or often unfairly dismiss Nature – The Greatest Master of it all. Written out, that's 40,000,000,000,000 of self-operated magic per person without considering all other species. If we were to create a better world and better system, such architectural design could settle our indifferences once and for all.

The amount of wonder, precision, and thoughtfulness that has indeed been put into the complexity of all life forms is truly magical. The breath-taking factor goes beyond a global raindrop, with a bigger purpose than we can imagine. Human engineering was skilfully woven into the Nature of living matter. Masterfully thought of and designed with dietary arrangements of "supply and demand" to reciprocate life force energy among all species avoiding overuse.

Just like oceans are made of many drops, and our bodies are made of many cells, it is a glue of unity, consideration, and oneness, showing its majestic greatness and presence in the world. Get ready to move into the Era of internal innovation and the possibilities to achieve the collective power hub, creating bandwidth potential via a unified global field. Natural Law states that the old must go to free space for the new.

When intelligent species live their lives without having the time or ability to apply the required essentials, we must ask questions. Who does the system serve and how to pivot? The essentials, such as affordability of organically grown foods, having time to cook a healthy meal, time to practise breathing, forming healthy families, watching children grow, hearing birds, or pausing to watch the clouds. Instead of being exploited by the system, struggling with ill health, and fearing the future that is driven by the sole purpose of accumulating resources.

Interestingly, from nature's perspective, the current state of the economic system does display itself in increased but normalised cases of obesity that display its pathogenic influence not only in the economic system but also in human health. The example is not intended to discriminate against people who struggle with weight, but to provide an analogy for understanding so one can attempt to pivot. Each body is born with a number of fat cells. Fat cells become bigger or smaller according to the accumulation and distribution of resources. In our economy, it represents resources, financial systems, and markets with its job of intermediate to provide equal distribution and regulate the exchange without accumulation except for emergency; interest rates and taxation represent life force energy produced by the healthy cells utilising resources to sustain life.

The higher interest rates and taxation, the lower life force energy is, and the more life force energy is wasted. Healthy people with non-pathogenic behaviours, lifestyles, utilising only what is needed, consciously and considering away of living, and awake, represent healthy working cells; unhealthy people in this model represent corrupted cancer alike cells that are not at fault but "effect" of a "cause" risen from artificially created environment. Perfect example would be farmed animals or birds oppose to those that came to life without restrictions, control or force. As species, we are meant to learn Nature's Architectural Laws, essential life skills to build foundation to step on, wonder and discover life for ourselves without force. Inadequate environment with outdated system that's out of sync with Nature's Architectural Laws are the causes of disability, illnesses, poor reproductive, mental and physical health including inability to see or understand peripheral perspective. It does not mean not smart. When population with none peripheral perspective become majority, are no longer able to see indirect pollution, wasteful consumerism, health harm and on leadership levels unconscious wasteful funding's. When supporting pillars of the Nature's Architectural Laws of creation are not at core for strengthening, and continuance of the community, it only accelerates decline.

Foundation of Balance is Precision. Just like accumulation of waste that contributes to the reckless diet, none NatalDNA or reluctance to change; eventually, will change inner body environment, which will store toxic waste or grow fat cells rather than cultivate clean life force energy. Such later serves as foundation for anxiety, mental health, cancer, tumours, inadequate understanding and pathogens and so on. Pathogenic cells in simple definition – are healthy cells that become carcinogenic or pathogenic due to the environment that is set by System, Diet and Lifestyle that are not in sync with Nature's Architectural laws. Consuming resources and energy for self or just to support own kind without limitations, instead of living in harmony supporting all life forms. By doing what right, taking only what needed and reciprocating back to Nature - by noticing with gratitude, admiration and caring for all life forms. Mechanics of reciprocation inbuilt within human design - Where attention goes energy flows. While, pathogenic cells or bacteria embodying themselves deeply, expanding within internal organs occupying all possible pathways for the sake of consumption and accumulation sitting tight caring only for itself and its growth, without reciprocation, distribution or contribution to Nature's Life Force or balance of the body with its hardworking healthy cells that keep our body a life. It's not about what we want, but what nature of our cells need to complete its best life span. To the point that the hard-working cellular system of the healthy cells can no longer curry, its weight accumulated through ignorance and greed, struggling to sustain it or shake it off, leaves nothing but driving the full system set up into fast decline.

When fat cells get bigger, wealthier, and ballooned to be carried and sustained by healthy body cells storing valuable resources rather than allowing healthy cells to utilise them, the balance tips driving mass exhaustion of life force energy in the healthy cells, reducing its life span, and causing the system laps. Representing complete loss of balance, where it consumes life force energy at its highest, and where essential body cells are no longer capable of existing. Our contribution to our planet environment as species has the same effect. Therefore, aim for learning, understanding and acting upon peripheral perspective is a must.

If we are to explore tapestry waved into nature's circle of life, everything else becomes simple. The Nature of our body and all living matter works on

7

elements, precision, energy, consistency, frequency, and power. From nature's perspective the circle of life is defined by the precision in resources that are required to fuel life force energy and life force energy is required to fuel all life. Both are most fundamental, but only effective in a perfect effortless flow of balance. In the fundamental laws of nature bandwidth of flow of unified energy, high speed of communication and innovation among cells can be only restricted or limited, when the human body as well as the global system overload corrupted with waste, corrosion and toxicity that originates from unsuitable or trapped resources, environment and food-induced elements that are toxic to its life, just like overloaded highways become congested opposed to congestion-free road when fewer cars are about. Or an example of an overloaded aeroplane that's unable to lift off. Many food innovations and tech prevent human faculties, abilities and talents develop further for the same reasons.

Just like we plan school year curriculums for our children each year, nature does the same for us too. Only in nature, it happens over each century, where she carefully observes our development and advancements. Why would nature want humans to achieve evolutionary advancement, and precision in the circle of life, you may ask? Well, precision brings elemental food allocation, which helps balance and facilitate exchange in production and reciprocation of life force energy among all life. In nature, human intelligence isn't defined by the level of IQ but by the ability to understand the purpose and diverse tapestry of life behind the reciprocation of life force energy, the importance of balance and what it must do to pivot towards the right direction. By simply knowing that it is the right thing to do. In Nature, such power is known as becoming or growing into a human being, acknowledging itself as part of humanity. Growing into, or becoming a human comes with achieving the level at which one understands nature's essence of life, thriving to achieve conscious unity, wisdom, emotional intelligence, and activation of extra sensory perception (ESP), including reciprocation abilities to self-sustain and support all Life.

Such energy, especially noticeable during prolonged fasting, one meal per 24 hours or minimum 18 hours fast. That's your direct plug-in or direct line to the Divine Life Force and Power of inner knowing. More so, intuition gets sharper by the day allowing to be guided from the heart. Without NatalDNA Diet the effects may take longer to develop and take effect. All depends on several factors we discuss in this book.

In Nature, self-sustained rechargeable batteries exist within the human body, which are required to function on limited resources yet produce high levels of life force energy, giving our body extensive mileage while working on a minimal fuel, including evolutionary health advancement and efficiency in function. Does it sound familiar? What's in it for Nature how it's relevant to our past, science, and industrialisation, NatalDNA precision, the food we eat, systems we create, and how it all has its purpose, we go into detail in this book. We look at the power of one and how we can pivot to take control, retrieve nature-given faculties, and health, and build the world we all want.

The ancient knowledge that has been passed onto humanity via many different pathways is the wisdom of Understanding Natural and Universal Laws as a core of Nature's Alchemy being an Essence and Nature as an Architect of life from cellular to grand being simply genius. Whereby, one can exercise a remarkable ability to view, see, and evolve outlook from different points of view and perspectives, knowing exactly what is right or wrong and how to change the course of actions individually for the best possible outcome collectively, knowing change is inevitable part of life and nothing that no longer working can remain the same. Reverting to the individual function of a healthy cell or a means of a raindrop contribution to the ocean being explicitly self-driven, equally important with unconditional trust in the laws of nature knowing nature has its best interest at its core.

How are we wondering off track and why we must do our part? Natural Laws and taboos within, are designed to protect the balance for all life forms, including you. Retrieving an understanding of Natural Laws will serve as guidance to achieve what we must. For instance, the manufacturing of materials that are harmful to life forms and that nature cannot recognize as its own design are incompatible and unable to break down internally or externally. Making pre-ordinated Natural processes and order internally or externally work "out of sync".

Nature's engineering is defined by simple rules highlighted in the following statement: "When something is gained - something is lost; when something is pulled - something is pushed. No one can be a master of Nature, but the manager and guardian of balance through the body it was given. Its power can be only borrowed in the form of intangible time in nature given body to experience life and do what it must, bound by Natural Laws. Everything we take or discard has its effect, which in the end, we will have to bear the outcome".

Anything out of balance disrupts the model flow. Just as we witness in the system, we created those fuel economy yet denies of balance. The effects of our choices are severing weather patterns, earthquakes, and volcanic eruptions, which are Nature's coping mechanisms. Just like too much sun will burn or too much water will cause floods. Our lack of understanding causes an overload of waste and toxicity beyond the capacity of nature's processing time will have its price. It's the effect of misuse of the power given to us by Nature and the reason many civilisations have fallen. When disposal happens faster than it can be adequately processed, it violates the Natural Law of Time, leaving no escape from the consequences. Will we become a civilization that will pivot and do what it must with guidance in this book? Or will we become the civilisation that falls too? The Law of Liberty or "The Law of Free Will" is given to the conscious mind within everyone to make that choice, because nothing in Nature is done by force.

However, the human body is an extension of the Earth, bound by the Laws of Nature, governed by the same Laws and the same principles and taboos are applicable within, despite anything we may think. Unity between the human body and conscious mind, known as embodiment, definitely plays a part. In short, it usually happens when an eternal entity becomes aware and considerate to humble and valuable service of not an eternal entity of living matter but rather bound by constant change, in the form of a body. And when there are over eight billion human bodies, where each doing what they want or what they feel is right for their own system ditching the natural laws, it rather a large number for Nature to balance and level up. What do we really know? Considering that we can't even get our systems in order and balance. Searching for a perfect solution by looking in the wrong places and fixating on something that is always changeable. Perhaps it's time to step back and change the angle of what we see and how we view it.

It is crucial to remember that all countries, land, oceans, and every cell in your body governed by Nature, belong to this planet, to the Great Mother, as their primary custodian. Therefore, her ways should always be our first choice.

We must strive to understand, live consciously, and comply with her ways, not to solely benefit ourselves by seeking ways to exploit resources for profit or just because we feel we can, but to find ways in which humanity can contribute to reducing our ecological footprints. Honouring our planet by embracing stewardship and Life Force Energy Conservation. We should actively work towards facilitating life by harnessing and conserving life force energy in all its forms while minimizing our reliance on manufactured and modified resources.

History, religious beliefs, indigenous knowledge of the land and fasting. We've eloquently touched on a crucial insight. Throughout history, conscious practises like fasting, harmonising with nature's rhythms, meditation, art of energy movements, performance and dance, gratitude, and communal gatherings have existed, often without complete comprehension. As society increasingly relies on scientific validation, driven by mere profits, our understanding of these fundamental practises and their underlying principles has gradually slipped away, leading to a profound detachment from our natural roots.

Nature has endured for centuries, predating our existence, and will persist for centuries beyond us. Its resilience remains steadfast, even in the face of our inability to connect with the inherent wisdom ingrained within us. Even if the majority of us succumb to ignorance and inaction, triggering a reset prompted by unbalanced forces, Nature will unwaveringly follow its laws.

The fundamentals of Nature often reside in common sense; however, many have become disconnected from this profound gift. We don't necessarily need to expand our economy, deplete the soil, increase food manufacturing, overfish, or harm more animals. Instead, our path lies in achieving balance in harmony with Natural Laws and understanding how Nature operates. After all, the nature of cells inside our body does not perform their daily chores for profit, but to sustain the balance of our body, and if they did, I doubt any of us would have the body to experience its life.

Reconnecting with the wisdom of Natural Laws, and comprehending their intended course and function allows us to take a step back and witness a different, more harmonious display. Perhaps we start with ourselves and let Nature correspond back to us with a magical display of what is possible. We only need to give it a chance. A positive initiation would be to alter our dietary habits to align with the natural diet bestowed upon us by Nature's design. We the last hundred years we had many experts, yet hospitals are full, and nature is dying. For sure we must ask why. Did nature create any species without being equipped or is it our addiction to control and meddling that made it happen? Embracing the sources without modifications, chemicals, additives, taste enhancers, flavourings, or meddling with foods outside the boundaries of natural laws is our most crucial way forward. Our nature-given body is equipped to work with nature-grown foods not factory-made. Trusting nature to guide us.

NatalDNA diet is not new, but rather ancient architectural design is woven into the tapestry of life and rules of Nature. It is the particular group of foods allocated to each organism matching its own unique genetic template by Nature at birth, adjusted with ancestral patterns, to facilitate the best building blocks, replication abilities, longevity, and cell quality of that particular organism, reducing deficiencies, inflammation, stress, illnesses and else contributing to fast corrosion effect and accelerated decline within the genetic template. Not without purpose to allocate resources as equally possible for all species to prevent exhaustion. Following Nature's narratives by achieving the NatalDNA precision in dietary food sources that are meant to work with nature's given genetic template and what will evolve from that, I cover further down in this book.

In further reading, I delve into Nature's given gift to all of us that can improve our health and shift our lives significantly, from hospital queues to a life of wellness, reduce waste, the need for unnecessary supply and unnecessary energy consumption, the need for extensive delivery logistics, waste and pollution by a minimum of 70%. Rather than focusing solely on renewables and recycling, we focus on ourselves and what Nature gave us, which at the end becomes a preventative measure driving "domino effect" of self-rectification for everything else.

When we focus on restoring Nature's given manual, we are pointing our life force energy to reduce our addiction to suffering, with a never-ending appetite for foods that poison our cells, depress our minds, and need for consumerism and scrolling, we receive so much more. The life force energy that is wasted on something out there and not the power that resides within each of you here, in your body, is an absolute waste of the power each of you holds. Our evolution lies in restoring balance as intended by nature, accommodating, and benefiting all life forms. This approach aligns with the wisdom bestowed upon us and is in harmony with The Law of Time, allowing us to manifest equal distribution of joy, wealth, resources, and equal life-work balance to accommodate all lifeforms, with you in control.

13

2 CHAPTER

THE WAY OF NATURE

The plethora of manufactured foods, plastics, chemicals, expired medications, disposables, recyclables, and non-recyclables disposed of daily contributes to a reduction in the lifespan for all of us, accelerating ageing and displaying early illnesses in our children. This is not a mere assumption; it aligns with a fundamental law of nature: fast growth often leads to fast decline, just as travelling at a higher speed gets you to your destination more quickly. Likewise, excessive waste brings about more toxins and pollution, making it challenging to eliminate. Hence, the key lies in precision and balance, according to Nature's invaluable laws.

Although individual perceptions of balance may differ, it underscores the necessity for a broader observational perspective. Remember, we are all unique but equally essential to nature. Developing an observatory perspective and understanding at our own pace is crucial. As the saying goes, a person becomes truly wise when they navigate through life as if walking on water, leaving as little waste, footprints, or damage as possible. This state of wisdom is achieved when one no longer craves unnecessary wants, focuses only on necessities, and maintains a joyful state of heart and mind by simply being a guardian of life.

Interestingly, the principle of "less is more" holds the true definition of terms of consumption applicable to many aspects of life. The law of cause-and-effect mirrors the economic law we've created—supply and demand. Eliminating unnecessary demands will reduce supply, and by removing the cause, we pave the way for different and more positive effects.

Although Nature has given us the freedom of choice, it will always retain its order to sustain the balance of life and support all living matters. Yet, gently guiding us towards our purpose, safeguarding us, just like we safeguard our toddlers, when they first stand or start their wobbly walk.

The timeless essence of nature stands as a testament to its enduring power and the interconnected wisdom woven into its fabric. Despite our transient presence, Nature's fundamental laws persist, offering a profound lesson in resilience and the cyclical patterns of existence. It is a reminder that, in our journey through time, aligning with the wisdom inherent in Nature ensures not only our survival but a harmonious coexistence with the forces that have shaped our world for aeons.

Consciousness, for instance, isn't just the state of being aware of and responsive to one's surroundings. In nature, it is a source of communication, which is present in and among all living matters. Nature communicates with all of us, responds to prayers, depth and pattern of thoughts, affirmations, guidance, and all nonverbal communication among all living matters daily. It is important to know the information one should allow in or reject. Thought forms our perception, understanding, and choices leading to guidance and actions. Just like the quality of ingredients in foods matters, each thought is designed to be mastered, to avoid waste of energy, and to attain the best possible guidance from the divine, leading to choices, actions, and quality of life. When one asks the question and is attuned to listening, the answer would come, and communication would establish itself effortlessly by the power of clear knowing. All about energy, flow, expansion, soul origin, spirit, the power of unity consciousness and its relevance to our diet, we will be discussing in the upcoming book "The Alchemy of Energy and its Infinite Power" in the Web of Life.

I am forever grateful for being so lucky to spend many of my summers at Nana's farm, made it easy to understand the preordination of nature's architectural design, the alchemy of life, easy communicative guidance and inseparable unity waved into the deeper tapestry of all living matter yet bound by Natural Laws. Imagine you have been given a huge piece of land with many pairs of species to support, knowing your life is dependent on every single type of species because each has its unique place and contribution to the wellness and longevity among all including you and your family. Nothing from industrialization or manufacturing or technologies.

Just you as nature's creator and representative, land keeper, and guardian of species, that been given a choice to come up with the best decision on the best possible food strategy that's sustainable for all. Knowing that each kind of species you lose would remove ten years of your lifespan. What would you do? There are no shops or supply chains. How would you manage? Naturally, if every species would eat just greens and grass, soon there be no grass left. Nature grows everything according to its life cycles. If all species were carnivores, soon there be no species left, because they would eat each other faster than they can be replaced. If everyone ate just fish or birds, soon there be no fish or birds left. If everyone's diet were some fruits, there be no fruit left for nature's needs to replenish its own strength, sustain the environment and so on. The wisest choice would always be dietary allocation that would serve the best health of each unique nature given the environment of those species. Will it not? Will it not prevent resource exhaustion, and accumulation of toxicity and promote last? And how would you manage, should your ancestry family have moved in with you? Remember there are no bin waste collections either.

Analogies used in this book are not to mock or undermine humanity but rather to highlight where we go wrong by expanding worldview from nature's perspective and source of creation. No matter how much we manipulate and interfere, the natural and universal laws inevitably assert themselves, operating on both small and large scales. We are challenged to create systems, yet these very laws may cause them to crumble until the right balance is found. On this planet, nothing truly belongs to us except the time we must experience its magic and beauty, sharing our talents, creations, energy, and joy with all life.

Our souls align with the flowing power harnessed and executed by an executive team of trillions of cells and microorganisms within our bodies. Each cell is as vital as a raindrop contributing to the creation of the ocean, and our presence in this ever-evolving world is significant—a self-operating and advancing system striving for ultimate unity, self-sufficient sustainability, and continual operational upgrade.

Just as the body is an extension of the earth and bound by The Laws of Nature, each cell is an extension of the body and bound by the same laws. Therefore, any foods, external or internal modifications, cellular or not, inevitably interfere with the flow of life force energy applying the "handbrake effect" on the global flow of unity consciousness and function of life. Every cell is conscious, cultivates, harvests, and transmits energy, communicating its wellness and health status always collectively, expressing itself through signals like tiredness, exhaustion of energy, lack of sleep, depression, stress, mental or physical struggle, deficiencies, inflammation, illnesses or low spirits. Equally, it reflects and projects the state of the earth empathetically, forming a symbiotic connection. Just like cells collectively express the health of the body, the global population expresses the health of the planet. In the end, even most hidden system defects or distractive pathways become clear in Natural Laws and can be traced to the source for solutions. This intricate relationship might be challenging to fully grasp, akin to the dynamic between governments and their people, where the needs and well-being of cells or citizens must be met rather than controlled for optimal equilibrium.

Consider this: the nature that designed your genetic template at the time of conception is surely considered the best nutritional composition to harmonise with the environment you would later be born into. Somehow, we know it isn't anything mass manufactured, not fresh, massively farmed, modified, and excessively controlled by our muddling with soil and animal farming. Nor the need for such high volumes. Nature, the creator of all life forms on this planet, has meticulously thought through the requirements for your body to align with its fundamental laws, sustaining the best health pathways.

The team of trillion cells operates in unity with each other, mirroring the intended harmony meant to be resembled within humanity. Nature's laws and order apply universally to all life forms, big or small. Since every life form, regardless of size, possesses internal cell memory to replicate and alchemy to navigate its existence, it is only reasonable to acknowledge that cells, the building blocks of life, are no exception.

The intelligence embedded within each cell is extraordinary — acting as a conscious creator, record keeper, connector, transmitter, and communicator. Stores memories of interconnectedness among all life forms and memories of our ancestors for endless generations back to our origin. Isn't it incredible, Right? The collective Cellular Intelligence of trillions within our bodies surpasses the population of eight billion humans. Suddenly, Artificial Intelligence seems less advanced, yet we don't see it. We are indeed consciously or subconsciously always striving to imitate nature. As we are fascinated with our own creation more than we do with the creation of others, be it children, achievements or tech toys, we are missing out on a living mastery showcasing its majestic power before our very eyes. The challenge arises when such advancements are exploited for personal gain, disregarding the Natural Laws.

Cells and cellular intelligence - are intricately linked with the Life Force of Nature and all life forms; each cell is inherently perfect upon our arrival in this world. We are, from birth, capable chemists, and architects of our own lives until we subject our sophisticated cell colonies to a lifetime of servitude and the dictatorship of the ego mind, behaving as if we "know it all." Indiscriminate consumption without questioning the impact on our physical and mental well-being fast-forwards cellular decline, straying from the original fuel source designated by nature at birth.

Much like we dispose of items, knowing they will return to the soil, we unconsciously treat our bodies similarly. We rely on authorities to regulate, yet they too lack the qualifications for such a subject—only Nature holds that expertise. Mass production, resource depletion, and chemical application on this scale have never occurred in the planet's history. If it took 80 years to discover the poisonous and addictive nature of elements in cigarettes, overuse of antibiotics or implications of wheat modifications, how long until we uncover the effects of everything else? Research is valuable, but its findings are often swiftly mutable as nature is ever-changing.

Even Artificial Intelligence doesn't update as rapidly as nature, whose accuracy remains questionable. In a world where online portrayals differ from reality; Artificial Intelligence might eventually shape our perception of ourselves and the world. Nature, however, guides us individually to unleash our best talents in alignment with our human and grand design. An analogy from an Indiana Jones film comes to mind, where an overload of knowledge leads to burnout—perhaps understanding of unveiling knowledge selectively via natural pathways is the key. The power of inner knowing only comes through pathways of clean energy life force that fuels it, opening understanding from peripheral perspective while allowing that life form to zoom in and out as and when needed. Just as each cell knows itself as a mini creator, working in harmony with others and nature defines the fundamentals of life.

.

3 CHAPTER

NATURE AS AN ARCHITECT
AND THE POWER OF PRECISION

We all heard of balance. However, many do not know that fundamental of Balance is - Precision. Nature uses precision in its Architectural Design of all thoughtfully created eco-system to accommodate and sustain all life. Outbalancing humans' interference is particularly challenging for such perfect engineering. We apply the same in our innovations, yet look dismissively on precision in all life forms and created around us.

There is no doubt that humankind are remarkable species, especially in understanding quality resources and precision in building and manufacturing processes. Let's look at this phenomenal analogy. Sand, for instance, is the most-used substance in the world after water. But did you know? Human civilization, as we know it, could not exist without sand. It is used to make concrete, glass, roads, and houses that require tonnes of sand and gravel to build. More importantly, the Sahara Desert sand, which is formed by wind, making grains smooth and round, isn't suitable. We need the sand that is rough with jagged edges that come from beaches and riverbeds. Such sand is formed by water. The very reason why the world's tallest building, the Burj Khalifa, surrounded by desert, had to import tonnes of suitable sand from Australia. If inappropriate sand was used to make concrete for the building, it would have collapsed.

The point is that the precision of elements and their influence on end results play a crucial part in everything we build. Each life form, including the human body, has been designed to work on principles of exact precision derived from foods. Since Natural laws are an inseparable part of Nature's design and are an inseparable part of nature's tapestry of life within all living matter, its design, and laws too woven into cellular intelligence and an inseparable part of our body.

What it means is really simple: when we innovate by altering the nature of our foods or the nature of our body, the nature of our body cells innovates too, by using a completely different range of life force energy. The only concern is that because we are oblivious to Natural Laws, while we innovate wheat, grains, dairy and sugar, cells innovate into arthritis, diabetes, IBS, and celiac disease; while we innovate our diet, nature innovates into attention deficit hyperactivity disorder or other mental and physical conditions; while we innovate agriculture, overuse of meds, animal farming and warheads nature innovates into cancer cells, tumours, and infertilities. Therefore, it only makes sense to use its fundamentals by working on precision narratives within our diet as nature's beings if we are to succeed in repairing damages; especially since there are so many of us now for nature to manage.

Think for a moment about what is possible if we pivot and choose to work on the precision of natural laws and elements within the human body. The very reason why NatalDNA is our next and most important step forward in human evolution and the balanced sustainability of nature and all living matter on this planet is simply because, when one gathers skills, understanding, and wisdom to adopt such ways, all other needs and wants to evolve from there. Spiralling away those that no longer fit into the future and that are not in line with

Nature's building blocks of life.

Let us now explore nature as an architect, designer, and software engineer of health, elevated or fazed states of emotions, immune systems, longevity, sanity, and natural evolutionary advancements. Origin of balanced growth and disability of decline. We look at all relevant to nature's aim for precision in the balance between pathogenic and non-pathogenic life forms, including us as a species.

Imagine, if you will, a tiny seed nestled within the embrace of the earth. It stirs with life, bursting forth as roots unfurl, seeking the nourishing depths of the soil. With each passing day, this humble seed transforms into a thriving plant, daily gathering "handshakes" and collecting footprints from insects, bugs, and slags who are there doing their jobs in nature, local and not so local, helping the plant to build its own strength while embarking on a remarkable journey of growth, renewal, and interconnectedness.

As the plant stretches towards the sun, its leaves unfurl, reaching out to the warm embrace of golden rays seeking its introduction to the world. Each day, the sun's energy infuses the plant with vitality—an invisible dance of light and life. And with the arrival of raindrops, nature's delicate data on current global viruses and varieties of pathogens' upgrades infused into the plant's thirst is quenched, as if receiving a gentle caress retrieving the life source memory upgrades from the heavens, expanding the Force of Life.

But the story doesn't end there. In this vibrant garden, a bustling community thrives. Tiny insects and bugs, with their intricate designs and unique purposes, come together as collaborators in this grand symphony of life. Bees buzz from flower to flower, carrying golden pollen from one to the next, stopping for a rest on this very plant, imprinting, and ensuring the continuation of life, upgrade, and sustainability of immunity through their intricate footprints and diligent work left on that plant. Butterflies grace the air with their delicate wings, adding a touch of grace to this organic ballet.

In the midst of it all, caterpillars and worms weave their way through the soil, tirelessly tilling the earth and nourishing it with their humble presence. They leave behind their own footprints, contributing to the rich tapestry of life that supports the growth of this magnificent plant. And as the plant thrives, it becomes a sanctuary, offering shelter and sustenance to a myriad of creatures. This isn't written to take you through a poetic representation of Nature, as we all know it's truly remarkable, but to expand an understanding of what we became, deny ourselves, and don't see.

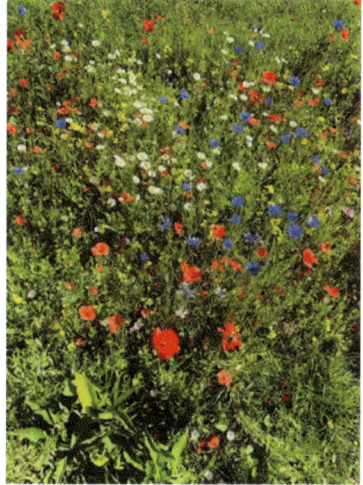

Nature's own seasonal manufacturing chain builds nature's own elemental, microbial, and medicinal resources for you and all species. May I point out that such complex exchange doesn't take place in non-traditional farming, and anything modified is fairly useless for the origin and nature of our bodies? If nature's bugs don't eat it, it means nature does not recognise it. As your body is an extension of the earth, the same applies to the body cells that sustain your life.

We continue. Nature's elements also lend their hands to this enchanting journey. Morning fog delicately blankets the leaves, weaving an ethereal aura around the plant. Dust particles carried by the wind settle gently, adding a touch of mystery, foreign touch, and intrigue. The rustling leaves, accompanied by the melodic chirping of birds and the whisper of the wind, create a symphony of nature's sounds, a mesmerising backdrop to this living masterpiece.

As the seasons pass, the plant continues its upward ascent, revealing delicate blossoms that beckon the world. The vibrant colours and intoxicating fragrance captivate not only our senses but also the creatures that have played a part in this botanical symphony. The pollinators, drawn by nature's invisible invitation, flutter from blossom to blossom, ensuring that the cycle of life and daily microbial transmission to reinforce our immunity and immunological upgrades for all life forms continue.

What does this all have to do with me? You may ask. The answer is: "Everything!" Naturally, when natural immunological cycles are not in place via our actions, we will be receiving a bare minimum of such power. Organically or widely grown vegetables, herbs and plants with nearby meadows will boost the highest upgrades in our immunity than anything else.

And then, the culmination of this wondrous journey arrives—the moment of harvest. Skilled hands pluck the mature fruits, vibrant and teeming with the energy of the sun, the touch of raindrops, and the collective contributions of all the life forms that have woven their stories with millions of bacterial footprints as upgrades through stewards of nature and regular raindrops into this very vegetable, fruit, or plant.

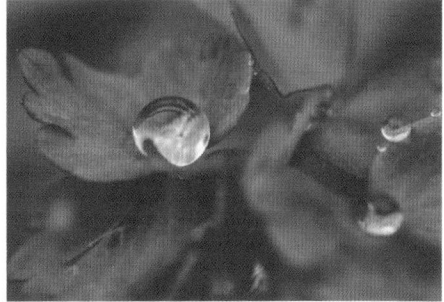

One can only imagine how many raindrops and how many times those raindrops went through the water cycle around the world, bringing the bacterial footprints, insects, and virus upgrades with them from the rainforest in the Amazon to the African waterfall, a local meadow, or a farm, landing on those greens, vegetables, berries, and fruit that we afterwards ingest. All this is preordained by nature, to integrate health among all life forms, day by day, drop by drop, inducing micro-doses of natural vaccines. Upgrading our cellular network based on what is happening with all life forms around the world keeps us and our team of trillion cells in sync with constant synchronicity and upgrades. Just imagine that thousands of bugs, butterflies, and bees that visit the vegetable you are to ingest, especially organically grown on local farms, are the highest quality nutritional and immunological upgrade you could ever receive. Can you imagine how few manmade vaccines we would need or if any at all? Now, imagine the same vegetable artificially farmed without all the stewards of nature and without regular rainwater upgrades with the bare minimum that goes into the fruit or vegetable, with the only benefit being profit.

Immunity is when the body has enough building material in the form of elements to sustain the function of essential systems within the body and defend it from external intrusions, while overgrowth of internal bacteria good or bad deprives it of essential nutrients required by healthy cells to function. Similar to the boarder control fleet.

When the natural order is in place, providing high precision and quality essentials for citizens in the form of people or cells, everything else falls into place, levels out, and bounces back, removing any form of instability within. Imagine a country that didn't have enough border forces because almost all fell ill or were called to deal with internal disorder.

The internal disorder happens when life essentials are jeopardized, and the understanding of interconnectedness and correct forms of communication among all living matter is broken. Although we need medicines for extreme cases, the body is designed to make its own elements to heal itself when the precision of essential matches is provided. Preferably without dominating garbage in it. Because even if it's useless, just like a car motor running without need, will still keep the cellular team working harder, draining life force energy, using backup resources meant for old age. Whether internal body function, a local county, a country, or the world, the model is the same. As insane, funny, or bizarre as it sounds, it all starts with what goes into a person's mouth because all wiring starts from there, with a never-ending building site.

4 CHAPTER

NATURE AS AN ALCHEMIST OF LIFE

"Alchemy of Nature's Elements" is living mastery and an intricate tapestry of life force, striving to make space for all living matter to live in unity and balance among each other and all life. While species and elements can complement each other, evolve, and bring out the best talents in each other to thrive, support, and sustain all life. However, there is also the opposite. And it's the load of opposites that throws the system out of balance.

Modification in Nature's eyes equals intentional cause for impotence, debilitation and undermining stand against its Mighty Power. Dismissing the nature of the origin and entanglements among all life of the living matter. It's not what we do, it's how we do things.

Moreover, to add fuel to the fire, we start to modify the original design of seeds, grains, fruits, and vegetables that have taken millions of years for nature to keep upgrading and develop in line with all life forms, only to transform them into something that nature of our bodies cannot recognise. And this is on top of chemicals we ingest, put in the soil, pollution, and waste we dispose of. If many medics spend years studying the influences of medications on the human body, one has to be super-mindful to prescribe such chemicals, knowing that a few little pills can cause harm to human health.

How is it okay to think that all additives and chemicals fed to birds and animals, added to our foods and soil, have nothing to do with the depletion of our health? Nature must be thinking that we are either dumb or blind because our cells can no longer recognise such forms, which are no longer compatible with the system that nature originally pre-designed. We were given all in pre-ordination, yet we can't live without poison. It's like someone handed us the most sophisticated spaceship, which we managed to turn into Kite, being proud of side streams of income we built for profits to use completely unsuitable materials to rebuild it again into the spaceship yet wondering why it doesn't work and goes down with any slight wind. Burning all resources into thin air and still proud of our achievements. Or analogy of being given a strong castle that stood for centuries, yet we knocked it down and built a crooked mixed plastic house in its place equally, being very proud.

In reality, nature as the alchemist of life does not produce a harvest just for us either but is doing what our body does — balancing the elements to sustain all life forms. It is time for us to learn and realise how it is done. Such modifications are no different to fitting a car tyre into a bicycle. Will it work? Or do we agree that everything must have its fit, precision, and purpose?

We sure thought it was only us who were that intelligent, right? This is relevant when foods are compatible with the NatalDNA template and are organically grown by individual and local farms. There is more. Nature collects data through the composition within the soil, such as insects, bugs, manure, water with waste, and animal and bird excreta, including human waste, and produces crops according to the level of health elements that species will need that season, including how many of each species; in accordance with what their job is and what their contribution is to the bigger picture, that we often do not see. It reminds me of food tasters, gourmands, reviewers, and critics who would check up on what was missing in a food recipe to suggest a resolution. Nature does the same. Each season gets

27

checked for elemental composition in nature's Garden of Eden, including what's missing and needed, how much poison is in the soil and in the air, what's changing, and how to neutralise it. Very much alike, like the human body, only bigger.

Nature's habitat of birds and sounds will add upgrades of sound and vibrations to upgrade and adjust a higher level of awareness, intuition, and the power of knowing to guide all habitats of life. It's like nature's own self-operating elemental global lab, architectural advancement, engineering, food production, and energy-generating centre all in one. Suddenly, AI isn't much of an advancement at all. But you are with that body of yours with its unseen advancements and cellular intelligence that needs fine-tuning, cleaning, and adjustment to nature's given fuel and ways, so it can guide you, opening a lifetime experience unlike anything else.

More so, not all harvests are meant for us. Nature will produce more greens, fruits, or vegetable types according to its collected data to balance elements for the health of the habitat and the soil. And not all harvests are meant to be taken. Some must be left for soil to serve as natural compost, and some must be used as food sources for other habitats of nature, such as birds, insects, bugs, and slags. Such a team builds nature's own vaccine in small droplets for all life forms. All is for us to share, and so when the time comes, we all, as creatures of nature, are set for the winter. Did you notice? Nature does many things in the ways we do, or, should I say, because we are nature, we have it built in us but claim it as our own.

I bet you never thought of such perfection and precision in nature that you thought of everything so well without human meddling. Daily clouds from all the countries in the world, repeatedly, just to support that growth cycle and sustain its strength, nutritional value, and life. Perhaps that is the reason that plants and vegetables grown by nature have indeed superpowers we never thought of because it's not just nourishment; it's the health, life force energy and immunity of global upgrades every season in all its glory and so much more. As we ingest it, our trillions of cells receive upgrades for the very season. It's unconditional love from Nature's all life forms, making it so extra special. And, when at last it makes its way onto our plate, without the need to overload, with each bite of that one vegetable, we become a part of this global life story, savouring the culmination of nature's artistry and collaboration. Because, it is savouring mediates the effect of nature on positive affect.

Did you know that all herbs, fruits, and vegetables are only really beneficial during the season of that year? Wild berries, nature's best vitamins, and herbs are the most precious and diligent healers. While some seasonal wild mushrooms, not farmed, especially locally, in very small quantities are part of the nutrient cycle, fungi upgrade within the body's local ecosystem to support effective digestion, gut health upgrade, and balance in preparation for the winter. These are just examples, and of course spices, plants and many more.

This journey, filled with wonder and curiosity, reveals the intricate tapestry of life that exists within a single vegetable and fruit. Let's not forget spring water. The true Queen of health and nature's direct line for the best connection, bringing cellular memory upgrades with every drop. It showcases the remarkable interplay between insects, raindrops, sunlight, fog, dust, caterpillars, worms, birds, and butterflies—the symphony of footprints and upgrades that occur with each stage of growth and the daily water cycle, filtered by nature's minerals and retrieved from local springs. It reminds us of the power and beauty of nature's renewal as life perpetuates and flourishes. Yet again, how smart we are to keep such goodness stored in plastic and mixed with micro plastic? Encouraging such practice in schools under pretence and wrong sense of health and safety. Building such habits in young lives that look up to us for looking after them. It is we, who distance ourselves from our true origin and true mother.

So, let us marvel at the wonders of this journey with each bite, embracing nature's most sophisticated seasonal upgrade to our immune system with the internal integration of global data with such precision into the human body of trillion cells and the dedication and interconnectedness that bind us all. As we savour the flavours of the harvest, let us cherish the footprints left by every creature, the contributions of every element, and the intricate dance that unfolds in nature's grand design. In doing so, let us wonder what part we played today in contributing to this intricate world of delicate cycles of life that contributes so much to ensuring the continual renewal and prosperity of our shared world.

While we lose ourselves in the sense of importance and entitlements without any contribution but slavery of nature, stripping resources at high speed and without limitations, and abusing the very life forms that quietly support us. Such behaviour would be intolerable and unacceptable in your children, should they go and spend all your life savings on self-glory toys within a year, will it not? What would be the cause for such behaviour? Understanding the Natural Laws and our role as a species. Because knowing and understanding are two different things. Our lack of understanding is the cause and effect of our behaviour as a species, ill health, and the world we created, but we can also be the change. Perhaps it's time to honour, understand, and protect this very life that does so much for us and define our actions by minimising our footprints.

Naturally, man-made goods that we dispose of make their way onto our plates. Man-made foods that we continue to modify and innovate do not get recognised by our cells as they have not been designed by nature to recognise such nonsense; vast farming and consumption of such foods do absolutely nothing for our health except our self-centered delusion of our grand knowing, malnutrition, internal exhaustion, anxiety, depression, mental health, and expensive medical treatments. Illusively convinced by fake looks, fake taste, fake representation of health, and fashion-driven stigma that all that food is supposed to be good for us, which is only followed by illnesses, infertility, frequent hospital visits with further cost to our health, and accelerated decline. Unless we fix our food, drinks, and lifestyle narratives to fix our ecosystem, any hacks will be short-lived making our system work harder and wasting valuable life force energy to clean up everything on our plate from decades ago, wondering where we went wrong.

This understanding challenges our conventional beliefs about the superiority of human intelligence. Nature's complex processes, such as the water system with its daily cloud formations from around the world, support growth cycles and sustain strength, balance, nutritional value, and life. As we marvel at nature's perfection and precision, we realize that our connection to nature extends beyond nourishment but an orchestrated effort to maintain health and Nature's intricate connection among all life forms.

Regrettably, we often treat our cars and houses better than our bodies and cells. Nature favours simplicity and has its own systems, as do our bodies. The Nature of living matter may temporarily accept modifications, but ultimately, it rejects what is not hers. Therefore, while longevity may improve, progression in disabilities, infertility, illnesses, and cancer will progress as inevitable for nature to get our attention, so we change our ways, including the way of thinking and patience that comes with it.

Human brilliance can be truly a gift and a curse. The gift is to recognise that its brilliance has its power of understanding to acknowledge when it has gone too far, there is no magic pill but only time to make change. Our actions, driven by a lack of understanding, contribute to ill health and environmental degradation. By minimizing our footprints, honouring, respecting, understanding, and protecting the intricate web of life. We can be the change.

5 CHAPTER

NATURE OF NATALDNA MICHANICS

The state of health can only be defined by perfect compatibility match among template and material used in the form of food given that life forms by nature at birth.

NatalDNA template, diet and biome master creators of unique life forms and a life form environment to sustain a healthy life. Is unique to an individual genetic template, with the best available genetic information from both parents that was available and preordained by Nature at the time of conception.

Nature of the conception, compatibility, and schedule. NatalDNA primary template would have higher compatibility with nature given diet and biome as opposed to the secondary template. Secondary is usually used later in life, when a swap is needed, either due to back up resources being depleted, or the replication template warning is out. It usually happens when we start to wonder why we have certain symptoms of foods we didn't have before or health concerns that weren't there before. Comes with comments: "I never used to have that effect before or I've always been ok with those foods, or such nonsense is impossible it can be that." Well, the nature comment would be the same as the comments from your car mechanic. It is worn out! Consider yourself lucky, because if the backup template may not fit as well, it's still working and can extend the lifespan in your body for the time being. Just as a secondary car part being a better choice than none at all.

Conception happens when body fertility, environment, and nourishment are adequate to sustain life. The environment must be favourable for a genetic contribution from both parents to work with each other and fit perfectly, like a jigsaw, including a backup template. During nine months of pregnancy, each system template selection is built in synchronicity with the environments the new life would be born into, according to the environmental data received from the mother's plasma. Towards the end of pregnancy, the strongest template set is more or less established. Life could exit early if the template became father dominant and mothers' diet, biome and environment are a struggle and no longer suitable to complete the last stage of growth. Hence premature birth. New life NatalDNA template compatibility would do its best to adapt to elements within mothers' plasma that originate from foods the mother consumes in the form of availability at the time of conception and its environment until the first heart bit. Further, adaptability from minimal to maximum or most complex elemental composition happens during nine months of growth and up until one becomes a young adult. Plasma and its environment are created by the body's manufacturing process from foods we consume daily. To start life will have its minimal requirements. To continue, nature will attempt to work on its best available template even, if at a minimal capacity, and adopt and compensate where possible.

The best possible combination of both parents, the best available fit, fused into one working unit that is compatible with the mother's environment, undergoes growth, adaptability, and adjustment to the outer environment way before birth. Selected templates, primary and secondary (backup), must be capable with abilities to bounce off each other, swapping, and changing if they must, according to the building material in the form of food, the environment in which it operates, or the change in the environment that occurs due to the incompatibility of food sources with the NatalDNA template, biome and the environment in which it will be born. For one purpose only, it is to sustain life. It is a preordained by nature genetically assembled template in the form of a complete jigsaw with the ability to work together to sustain that life. The NatalDNA template undergoes testing, with adjustments made to the template where needed during the nine months in the womb.

Testing the external environment via mother-food habits, adopting its bacterial foundation, and a challenged immune system. In preparation for the environment, new life will be born into, and mother backup resources will be used at their highest. Especially, when the range of incompatible foods is driving the highest levels of toxicity causing an acidic environment in plasma. The state of health can only be defined by how compatible with the template the daily diet is. Nature uses a preordination process for all species to prevent the loss of environmental balance and resources, extinction, exhaustion, internal and external chaos, strong immunity to protect that life, and to keep in line with the Universal and Natural Laws.

When an architect creates the system, it gives time and the way it would work best in its nature-given pace, with nature's preordained individually allocated material, its individually compatible types, and quantities in the form of elements. The system then becomes well-functioned, creates, and sustains a well-balanced and well-maintained environment with minimal use of resources, recharged and readily available to communicate, respond and work with all body systems at its best possible capacity.

Each parent represents different environments, which not only differentiate between people's inner environment but also environments inside the body's organs and systems. When compromised with failed equilibrium, chaos follows. Imagine all mixed without order, destabilising the order of function and environment. Analogy and nature of the body working in perfect balance and what happens when it does not? Just as we will not treat liver failure with medication for kidney failure or overall other health problems, we should not treat dietary requirements, herbs, elements, and all nature's aiders as one fits all. We have already made that mistake with the overuse of antibiotics, adding even more mutation and confusion to the nature of bacterial habitat; that has its balancing forces, too.

Wrong foods including manufactures snacks, just as wrong petrol in the car or incompatible herbs, spices or supplements breed the wrong bacterial biome, creating a change in that body environment; making the borders within the cellular membrane as well as the connective border between body systems disrupted, compromised, or broken, affecting its function. One will find that once diet is in place, the need for antibiotics and possibly other meds will become minimal or not needed at all. Reducing the need to use taxpayer money to fund industries related to it, making funding available for earlier pension, free childcare, or else, and support holistic way of living. Just like all species, nature designed our body with everything it needs.

Making its rules very simple: freshly prepared, as natural, and organic as possible, mindful of nature's pharmacy, herbs, and spices as they don't suit all and each template would have its own match, fast daily as much as you can, avoid anything modified, use breathing exercises or/and high-intensity training, contrast showers, explore the art of dance, learn your own nature given breezing technique, spent time in nature and drink clean water. All these will connect with nature's pathways via your inner mass of cellular intelligence, earth energy field and consciousness of all living matter that will guide you and grow your comprehension, cultivating higher energy reserves and advancements in higher intuition, decisiveness, talents, unveiling evolutionary superhuman abilities, HSP and so on.

On occasion, foreign herbs or spices serve as immune boosters and foreign foods as awareness, yet regularity of such overrides its purpose and tilts bacterial balance, causing unnecessary bacterial expansion and the need for higher mineral resources. Did you know that soil with high bacterial diversity usually lacks minerals? All species in nature have specific diets and, as a result, a specific biome that facilitates the intercellular life of that species. The human gut and its bacterial biome work just as a bacterial pool in the soil with one bacterial community benefiting one geographical location and its environment, while another will be favoured in another location. The NatalDNA biome of beneficial bacterial pool replenishing and regulating reserves are designed for the NatalDNA diet of that body, and that body beneficial organisms are stored in the appendix, contributing to the re-enforcement of the immune function of the body. A lot like beneficial bacteria in mother breast milk that reinforces and stores reserves of beneficial bacteria that correspond and will aid in digestion factors for that offspring later in life.

Nature reads and upgrades it all selectively via architectural mechanics and according to the species it must support. Bacterial compatibility will also determine which plants will grow and last or which will wither. In the gut, the bacterial biome is allocated according to the NatalDNA diet to facilitate healthy replication of the template. While some species have a particular ratio of bacteria that will benefit them, others will have a completely different display. When the NatalDNA diet is followed, the bacterial biome that is not beneficial withers away into dormancy, allowing nourishment to repair damaged areas and thrive. However, the moment a nonbeneficial diet is consumed, it will feed nonbeneficial bacteria in that body to thrive, robbing cells of essential nutrients, changing the environment, and allowing the parasitic side of nature to thrive.

In most cases, it isn't about the health condition but rather about internal bacteria that feeds itself into that condition making a home for itself, and the body owner unknowingly or arrogantly confidents that it isn't anything to do with its daily snacks, drinks, sweets, and food habits, feeds such growth. Depriving healthy cells of minimum nutrients for function and self-repair. In some way is no different, how corrosion feeds itself into the metal creating

36

rust, when the environment is favourable for its growth.

Further design extensions of immunity and its natural upgrades system. There are two types of immunity: local and global. Both works daily and simultaneously. Global pathways are the water system, rivers, winds, rains, and the oceans, i.e. cell and cell memory, water and water memory, bacteria, and all forms of insects, plants, and viruses. Even waste from living matter is the most important as it leaves its minuscular imprints carried with water systems and clouds. Local pathways are locally grown foods which empower the body with Nature's local bacterial upgrades of that land. Foods grown in other regions are naturally okay, if there is no other choice but for ideal health and immunity, foods from local land have power over others.

Nature was always designed to work on its own waste and recycling without confusion, until manmade waste, drove overload driving bacterial hybrid. Due to the tilt in balance over the last 100 years, self-operating and self-upgrading systems work out of sync. Just like having two different software running on your computer trying to compete and fit with each other to do the same tasks. Clashing and overriding each other were overall neither 100% effective.

The highest possible quality of NatalDNA template replication can be achieved by the highest precision in NatalDNA food compatibility with the NatalDNA template, allowing pro-life microbiome to thrive. Just like perfectly built wall requires quality and precision, body does too. It takes approximately seven years for the body to fully rebuild itself, should one want the same body with new settings. While at the moment quality of NatalDNA templates is on the decline, with the pivoting it can change into the opposite effect. This is subjective to the depth of damage done and the body's natural abilities and backup resources to heal itself. Worth mentioning that while the soil is exhausted, many are pushed to supplementation to replenish what the soil can't provide. Simply, because we are feeding not only a vast population of humans but the vast population of bacterial dis-balance in human bodies feeding of the cells, soil, and water and stored manufactured foods. It is an overeating mutation in bacterial soup, amino acids and proteins, putting the system into delaying its evolutionary potential.

NatalDNA elements are the foods that are compatible with an individual NatalDNA template as well as each other. The team of trillion cells has never been considered in the context of mass and collective work. When one can get one's head around the concept, understanding its place and work involved in your body management system and function daily is rather empowering.

Especially when translated into our manufacturing system context. Knowing what is possible with 3000 employees gives us an understanding of what's involved with cellular mass capacity that is 4.5 times larger than people on this planet. Suddenly, putting everything into a different perspective. Such mass is fuelled with abilities to thrive and keep your body alive or to perish. Particularly, when it comes to the natural law of oneness, reciprocation, and consideration of giving and receiving among all cells and the nature of all living matters. Power is always bound by the law of free will but based on intelligence and understanding, directing the action of preservation or distraction.

Curiosity has always taken me to the land of wonder, which I would like to take you, so you can wonder too. When nature creates changes by creating genetic variant, it is for the purpose of adapting to the changes in the living environment by using the same gene differently, working at nature's slow pace. Just like we adapt by either using the same product for different applications or altering product composition to the best fit.

Nature would do so by only altering a single amino acid in the protein product, in a space of time, when foods consumes do not undergo any major compositional change. Means the same food groups are without large variants of ingredients. The nature of the connection between the gene and the internal and external environment is that the same gene will work differently: by altering protein genetic make-up in one, while altering its expression in another. The only difference is that nature would only alter such with the nature of the environment to facilitate adaptation for species in need and over a long time. Making some health conditions being dangerous by living in one environment and perfectly complementing health while living in the other.

When we bring a wild animal from Africa or Antarctica to Europe, such a life will never fully adapt, not for several generations, but the nature of that genome would attempt to do its best. That is, without altering the precision of nature's given diet. However, when both environments are way away from the original habitat, the diet is out of sync with nature's precise narratives that have been allocated to that species-specific genome and without bacterial support of the original organic habitat permanently, fast mutations into extreme behaviours, inabilities, illnesses and decline are inevitable. Fasting has been threaded in all religions and cultures of the world for centuries for one purpose only, which is to prevent overload; that contributes to fast changes in the genome leading to unstable built and exhaustion of resources and soil due to over consumption.

That is before we begun to modify nature, foods and loaded our bodies with so-called variety. With all that slowly over 60 years changing our body environment. Became harder for the Nature of our design to keep up. Which not necessarily affect some, but slowly become detrimental to others, effecting children too. Nature created us all differently. We are just like organ community cells in the body. Some belong to organs that can work on one group of foods only, others can make and adapt to some variety, and others must work on absolute precision, being sensitive to change. No two bodies are the same. It possesses multiple pathways to address imbalances or pathogens as straightforward as a winter cold. While some of us can easily brush it off, others may struggle and suffer greatly from it. Nature uses the same allocation to all species.

Especially, with too many alterations due to mass variety and not a nature-created novelty with constant adaptation to internal and external environment with foods outside of the originally allocated diet for that life - is too much to deal with and too fast. When we participate in genetic alterations in foods or else, from the perspective of our own narratives, we nowhere near have the same abilities as nature, nor have we created nature to be equipped or qualified to do so. Because there is simply so much more. The same alteration would work differently when part of one template rather than the other in response to the environment and if the NatalDNA diet is considerably changed. Implications of this may not be visible until the third generation onwards.

While the science is puzzled about the origin and function of building blocks stating: "In this era of extensive genome sequencing, many new protein families have been discovered whose functions are unknown" (Protein function – Molecular Biology of the cell. ncbi.nlm.nih.gov) The question is: discovered because it didn't exist before and we forced mutation of original few, by meddling with our nature given diet, soil, microbial system exchange locally and globally, air, water, source of our foods and more? Or? I cannot even think of an alternative option. Isn't a mass variation in building blocks and

their unknown origin in one building that would bring instability to the structure and its last?

Nature is set in its pre-ordained function, and it will prioritise what is given or recognised. Before growing around what it does not recognise, isolating or safeguarding such instead. In other words, it would use bacterial protein as a building block as opposed to the manmade lab-growing protein that it does not have knowledge of or tools to break down or convert. Just like the soil would decompose and benefit from nutrient-dense vegetables faster than opposed to manmade foods. Because each would breed a different bacterial habitat. Naturally, quantities matter too.

Let's look at an analogy that may help to get a head around understanding and importance of the building blocks within the diet. We all know proteins, amino acids, enzymes, and other elements involved in building our body. Let's translate all this into an analogy of the building site and what goes into the brick-making process: bricks, cement, brick layering and building techniques. And how important precision, efficiency, synchronicity, and quality are in all this. Since industrialisation, food modifications and so on, instead of, let's say, three variants of proteins, there are over 25 with added mutated bacterial proteins that step in when they get the chance; and instead of 9 essential amino acids out of 22, there are many others. Once translated into building blocks, imagine a building that is built with over 25 different variants of building variants of material that may or may not be bricks, may or may not be compatible with each other, and instead of cement any other binding components used with a mass variety of brick layering techniques, and just as it comes. Picturing the building made from such building blocks, binding elements, and techniques, translated into the current structure of the human body in many, puts diet and human health into different perspectives. Would you buy such a building to live in it? Perhaps only if you knew you must rebuild it better before it collapses.

When historical buildings require renovation, an architect carefully examines and selects the best possible material with the closest match to the original building blocks to complete renovation. The same principle Nature would use if one to achieve precision in renovation with each uniquely designed human body. Naturally, it is a major inconvenience for the system we created to achieve profits, but that is the point where nature grants understanding and the choice of "free will" accept or reject it, as nothing in nature is done by force.

In nature's eyes just like in the eyes of the good architect, high quality and resilient building structure would require strong frame and quality building blocks. The NatalDNA frame is replicated with the minimum defects when the precision of NatalDNA foods compatibility is achieved. These are nature's required precursors for that authentically designed NatalDNA template.

This only means everything is our own doing all alone, we all collectively hold the power to life, health, longevity or destruction and death. We always have that choice. Most knowledge has been passed on from generation to generation in different shapes and forms, such as religious books, cultures,

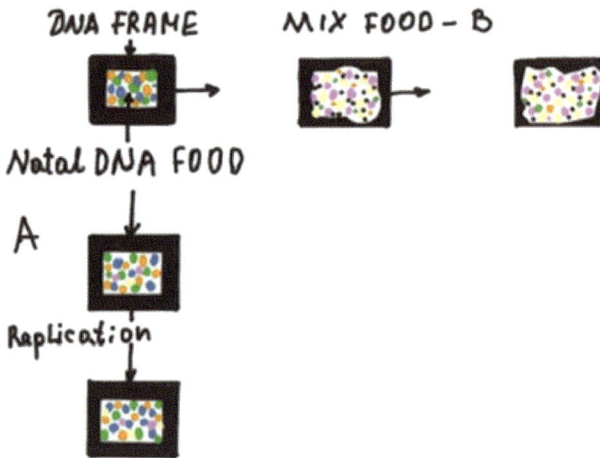

and traditions and many recognize the connection among such. The only difference is within the law of transformation the views of interpretation undergo upgrades to become relevant to the generations of that time.

On the image below A (on the left) representing the border well-kept is the display of the template replication effect with the NatalDNA diet compatible with the template. The other B (on the right) is an example where any food combinations are used. Reparative and sustainable results achieved with the best possible biochemistry between template and building blocks, inevitably become well maintained by the strong army of cells that in return continuously replace itself with even better-quality cells contributing to clear and then crystal-clear communication among itself and the outer environment updating itself and its pathways with Nature. Giving room for harvesting higher levels of energy by recycling leftover proteins and other locked nutrients organizing the perfect function of the house. At this state, Nature hopes that the human body evolves into becoming the highest source of energy generator, communicator, energy transmitter and more, with the potential to see the higher spectrum of light too. This is with the use of the least possible resources. For such efficiency, Nature has its plan, but in the beginning, it will develop self-rechargeable energy with abilities to rectify and heal inner damages. This is only possible through pivoting into a clean allocated NatalDNA diet, restoration of soil, and environment, and further evolution of practices. Such evolutionary progress falls in line with Nature's preordain design and as for the architect of life will be a dream come true. This answers our grand question: Why we are here? Knowing what we know now about the race for green, sustainable, rechargeable energy, it's easy to understand why Nature would attempt to thrive for the same.

6 CHAPTER

NATURE OF MALABSORPTION

When foods are unsuitable and incompatible with each other or for that body, those foods modify stomach pH, which then creates distorted communication within the body and stops food from breaking down properly, which then leads to a lack of conversion and malabsorption of required nutrients, deficiencies; and further need for food. In other words, if you are stressed, hungry, not sleeping well, having problems with your gut, digestion, skin, hair, brain fog and else - your diet does not suit you, making your body not function properly. The same goes for allergies.

When I realised that I can not be vegetarian without implications for my health. Finding hacks allowing me to use food sparely and minimally became very important to me. I also knew it was contributing to my future health. I then decided that if I found a way to hack my health without being hungry, yet eating a strictly essential amount, there would be fewer fish, plants, herbs, animals or birds have to give up their life to support mine. We are all interdependent and when we take from the land; we take from other species who share this life with us. When we reduce demand, we reduce the need for the life of another living matter to be taken to support ours.

It is the circle of life. Nevertheless, many life forms share this Earth, and we must use it sparely. And, when certain foods are not only useless to our health but also spike hunger to have another meal, which inevitably leads to taking another. Besides, it is up to us to see ahead and make the right choices by finding our own nature-given ways. By knowing that those ways must come at a lesser cost to yourself, other life forms and nature. When diet is wrong, we then have less space for the essence of water to work correctly to help us cultivate life force energy for us to heal ourselves and reciprocate, give back to the Divine, and grow. It is then an easy decision to know that for the welfare of other lives, removing such foods is the best choice for yourself, your children, and others. When such foods are contributors to malabsorption and are a waste of cells' energy by metabolising them, how does it benefit your body? Many of us are in search of understanding and must find their nature given the NatalDNA diet and our own water-drinking technique and breathing rhythm. It isn't easy, but it is worth it. For me, it was challenging to hack having one meal per day with no snacks or smoothies or anything except water, limited coffee and herbal teas. On some occasions, when I didn't sleep well or picked up a virus, I would need two meals. Appealingly, once I achieved such precision, I could never go back. Besides, my body wouldn't let me because it does not like any other way making me respect that. Getting such a level of communication with my body is already an achievement in itself. I am also aware that it will take the least seven years for me to see what it's capable of, but I'm ready to wait and continue to contribute.

We currently find ourselves in an era where food serves not only as a tool to earn a living but as a predominant means of social interaction. In this landscape, there is an abundance of manufactured foods that often outweigh the selection based on bio-authentic natural requirements. Unfortunately, inadequate food education and a lack of understanding of the intricate tapestry woven into nutrition and nourishment in all fields of life have contributed to a scenario where many foods are nutrient-deficient and laden with chemicals present in additives and the soil. The use of chemicals treated with chemicals often cancels out the very nature of the body's origin and function. Medical field study application of pharmacological drugs not nutrition, this must come from personal initiatives to understand. Just like a car mechanic shouldn't be responsible if you didn't learn and initiated to put the right oil in the car.

I make you another interesting analogy-based example of how and why in nature's eyes we are not the same, nor created the same.

This situation has far-reaching consequences, leading to numerous individuals experiencing conditions, deficiencies, malabsorption, and dysfunctions from before birth, starting at the time of conception. The interplay of environmental factors, nutritional choices, and the quality of food consumed during the critical period of conception and early development can set the stage for long-term health outcomes. Just as the incorrectly laid foundation of the building can drive its collapse earlier than opposed to slower and more resistant deterioration.

Addressing these challenges necessitates a comprehensive understanding of the holistic nature of nutrition and the vital role it plays in our well-being. It involves not only re-evaluating dietary choices but also promoting education on the bio-authentic natural requirements, personal responsibility and Natural Laws that support optimal health. Additionally, a shift towards sustainable and regenerative agricultural practices can play a crucial role in mitigating the impact of chemicals in the soil and promoting a healthier food supply.

The health conditions of parents, the quality of longevity genes, and the diet before conception all play a role, contributing to malabsorption and malnutrition from birth. These dysfunctions can lead to accelerated ageing (biological decline) or the accumulation of malfunctions that manifest as illnesses over the years. It's crucial to recognise that malabsorption significantly contributes to malnutrition and creates chaos in biological communication and function. Just as a clay effect in the soil makes it hard to absorb water or nutrients, wrong food compositions create the same effect too.

Even with a seemingly healthy diet, when malabsorption is present, the nutrients never reach their intended destination, resulting in cells starving. This process can lead to cells compensating by borrowing elements from elsewhere in the body where they are also needed, causing further deficiencies in other parts of the body or internal organs. Despite the appearance of an expensive organic and nutritious diet, the reality is that nutrients are not always effectively absorbed. Especially when not compatible with genetic templates, with each other or lacking conversion elements, turning what should be a nourishing experience into mere assumption and an expensive visit to the bathroom passing it through.

Understanding the nature of human design including precision in diet will address our self-inflicted cascade effect guiding us through essential steps toward helping the body to repair, promoting better health, and preventing the complications that arise from nutrient deficiencies and functional imbalances in the years to come, including the need for more food. Remember, your body converses in many different ways. Brain fog, tiredness, poor memory, anxiety, allergy, back or knee pain, indigestion, constipation, sleep struggle, headache, brittle nails, poor hair condition, skin pigmentation, cellulite or else, all health conditions besides the state of joy are forms of body communication with you, asking for help.

I recall me and my friend were fascinated when her hair would always turn from silky smooth to fuzzy the next day after she had a cake or sugar overload. It would become so fuzzy, making it impossible to brush. Yet, as soon as she resume to her diet, her vitamins and herbs that her body needs, within a day or two, the hair is back to silky smooth. I could not believe my eyes. Now, these are more visible with my friend and those who went through full recalibration and conversions to their nature-given NatalDNA diets. Such becomes a guiding factor because their bodies are no longer tolerable to rubbish. All it says is that my friend's body is very outspoken for the reason that 90% is functioning on clean fuel. The good thing is now or once she is older, if overloaded with any foods that she shouldn't have, her body will let her know and she will not be able to ignore it. Yes, it's not easy, but it's worth it. Considering that our body as it is, even on a clean diet, has to deal with all toxicity and pollution in the world that we have no control over. Until we start implementing changes. Giving ourselves the most nutritious food is the least we can do. Besides, each body's unique template would have different needs, meaning the quantity of required minerals and vitamins. Such detail would also fluctuate according to the inner environment of that body and the bacterial population that the unique body template feeds.

Next time you see someone losing their temper or struggling with mental or physical health, it is not more than caused by an unintentionally and unknowingly neglected shortage of required elements in the form of foods for years, finally manifesting its effects.

When the body is required to perform 30'000 functions a day, but the materials required for these functions were not supplied via the precision of diet that the body needs or has not been delivered to the cells due to problematic absorption or permeability defects, guess what will happen over time? Imagine what would happen if an electrician came to install all the electrics in the house, but only a few relevant materials arrived. Where is he supposed to get it from? Sending you an order through the pain in the muscle or a headache as the report for shortage, but unknowingly we often do nothing. Or imagine 30'000 jobs that sustain manufacturing complexes or energy supply systems were disabled or worked on limited capacity daily due to a lack of resources? Interestingly, just like a building does not show its cracks or defects immediately or soil does not display its exhaustion, the body does not showcase its malfunctions and shortage until later. It will continue dipping into backup resources to compensate.

Another interesting phenomenon is many of us would state that we have no wish to live till old age and may enjoy ourselves, yet when illness strikes, strangely hold on to life, occupy hospitals and would do anything to turn the clock back. As busy as we are, we still often do not comprehend that all the time in the world, all the busy schedules become available for the soul when there is body no more. Regret happens and is at the last minute and is irreversible. It is easy to think this way when the body's expiration date is far away until it knocks on the door.

Next time one has such an approach to life, take a deep breath and hold as long as you can and see how much that breath of life would be valuable to you, should you not be able to breathe anymore.

From the perspective of nature and human design, the human body must be closely compared to the soil because it will make and grow from what's been put in, changing its fertility, environment, and elemental composition accordingly. Just like soil can become infertile as an "empty well' due to exhaustion and lack of compatible elements, the body can, too. Such incompatibility of foods and drinks with the bio-authenticity of the body, as well as, compatibility with each other creates the "clay effect," which usually exhausts digestive enzymes and leads to malabsorption.

Imagine plastic or any other hard decomposable foods that are not Nature and are wrong for the soil to be put into the ground without any bacterial support to decompose. The only difference is that soil will keep incompatible foods for a while before being able to decompose them, while the gut will slide it through the system to remove it without converters, and as a result, there is very little use for it. While the body continues doing that, it uses up life force energy for no reason. Just like you going to work, exhaust yourself, add stress to yourself by receiving fake and useless resources in return, and use up your internal savings instead to support your health and life. Each human body is a daily building site and should be treated as such. Each building will have a different, carefully selected variable of material and the team that overlooks; the body does, too. Naturally, building from just any material will not facilitate the results; expect it to be a fortress. Think of that every time you put anything in your mouth, on your skin, on your hair, and so on. In the tapestry of life, every particle influences your existence on a minuscule level, building or destroying your beautiful body on a cellular level, one by one, from within, making you weaker or stronger depending on the decisions you make.

The best way to understand malabsorption is to think of water being poured into the leather, where it slides, and into the fabric, where it sinks, making a change in permeability. The same happens to too many with water. If there are no elements to take through the cell membrane, one can drink water all they like; it will go straight through them, leaving the body dehydrated. Try pouring water onto a clay and see what happens. We all know water is very important; it is also a facilitator, generator, conductor, builder, and transporter of energy among cells within the body, all living matter, the outside world, and the cosmos. Including everything known within neuroscience.

It isn't about the amount; it is about the quality of the water and the amount that is absorbable by the cellular membrane. Because spring water, for instance, carries system upgrades and life force energy, it deepens interconnectedness with nature and the divine powers that facilitate life. The inability to meditate, manifest, align intuition, discover life purpose, connect to divine guidance, and struggle through life, comes from that, as well as the mental capacity to understand. Some can understand but cannot switch on the core engine due to deficiencies, incorrect diet, and else is blocking the flow. Others can be aligned and guided intuitively from a young age, but without knowledge and understanding of the nature of living matter, they end up driving the system into exhaustion around their 30s, 40s or 50s. This often happens to life-achievers, who achieved success earlier in life, puzzling about what has happened to their confidence and feeling joyful about being high on life, creative streak, and manifesting abilities. Or what we labelled as mid-life depression or crisis. Left feeling down and pushing through life with the struggle of every morning. Guess what? The answer is elements and what is causing interferences in your body-building site.

Cells, bacteria, and elements can turn good or bad, order or disorder, life, or distraction, all depending on the environment they are in. While the sup of elements shapes the environment with the output of reactions it creates, the environment strives to keep order among those reactions pairing them away from disruptive and destructive behaviours in the best possible ways. However, if the input is daily doses of reasonable dynamite without any elements that will cancel its destructive tendency, please do guess what will happen over time. And this is exactly what we have achieved in terms of food modification and innovations. Order and quality of amino acids will be in disorder creating cascade effects into building poor levels of proteins that create poor binding properties with other vitamins and minerals building a colourful picture of cause and effect. On the building site, each grain of sand that the brick is made of is equally important as conversion and binding factors, brick layers and the environment it built. Just like it's impossible to build anything on a volcano that is about to erupt.

While, from the perspective of living matter, the system becomes polluted, producing energy reserves that are only left to function for the essentials, causing stagnation. "Bright Star" is only bright when internal particles remain in the same formula or overpower any particles that can interfere with the ability to retain continuity in creating the same fusion and retaining the required environment. Making itself self-rechargeable, like the sun, but without burning the living matter, yet recycling. The place and order where the very gift of nature's longevity is hidden.

The nature of living matter will always strive to achieve the highest sources of energy with the least possible resources to keep the balance. Because one does not exist without another. I hope you can see a clear pathway, interconnections, and entanglements among cellular health, nature's given diet, and the massive internal power that is given, often dormant, wasted, or misused within us. We are powerful beings on the soul level, but it is embodiment on the cellular level that ignites the energy potential within your body's energy-harvesting powerhouse, which is polluted and clogged. Just like a fertile environment within the soil creates magic for growth when fused with the energy and the tapestry of life, built on nature's focused precision, the nature of what our body needs requires such precision too.

We are the only species that voluntarily became similar to "lab rats "subjecting ourselves and our children to human trials and what has been named as human greatness. We are the only species on the planet, since industrialization, modernization, and innovation, have hardly kept our primordial diet, making the balance within the nature of our body and the balance within the nature of the planet tip. In nature, any destruction pushes forward transformative action. That means we need to go through and understand the nature of the body's origin and what it is built from to learn that we become what we eat. Weak or powerful; healthy, or ill; intelligent, or wise; mental, or physical – all have the same origin, we simply become daily from what we eat. How far it's from nature's intended genius is determined by the overuse of chemicals, metals, food manufacturing, inorganic farming, lifestyle, radiation, and waste we produce will have a declining effect on our health. Just as a nourished plant thrives without limitations, malnourished is limited to what it can see possible due to limitations in its building blocks faculties that originate from one's diet. Precision in nourishment throughout life opens and expands worldview in all its colours just like the same painting highlights and brings forward most hidden patterns and shades that were not visible at first. A gentle and attentive approach to each form of nourishment whether it be tea, vegetable, herb, fruit, or spice is an absolute must to achieve precision that works for you.

I will never stop mentioning that the overload of anything good is equally destructive as anything bad. Many elements cancel each other as well as complement; such reactions will be different for everybody due to the uniqueness of the environment that the body creates.

Nature's architectural design and essence are the presence of cellular intelligence and the power of the collective, not only in the unity of consciousness but in unity on a cellular level and for all beings within all living matter.

Nature has provided an abundance of all that we need as a species. Since we were born, all our focus has been on the system we created, with the constant

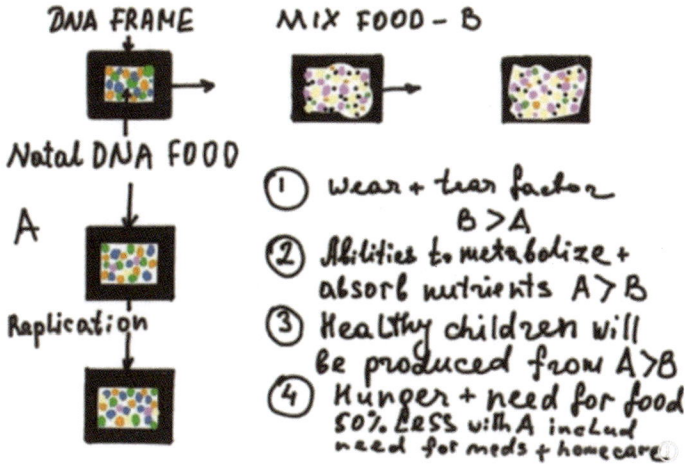

DNA FRAME   **MIX FOOD - B**

Natal DNA FOOD

A

Replication

① Wear + tear factor B > A

② Abilities to metabolize + absorb nutrients A > B

③ Healthy children will be produced from A > B

④ Hunger + need for food 50% less with A includ need for meds + homecare

unhealthy need to put our own mark on everything, innovate, and reinvent the wheel in a continuous attempt to conquer and control nature instead of harmoniously co-existing with it. We live in a world of systems, experts, and academics, yet human health and health in all species are declining. If we are supposed to be experts in everything, why is it happening? Isn't it supposed to be the opposite since we know it all?

Being simply grateful for what we've learned allows us to pivot and set a course towards greater things we can achieve by working with the balance of life and by innovating ourselves and our ways of thinking. Particularly now that we have nature's manual to guide us. Although it is important to know how body mechanics works, it is also important to know that there is so much more than that; two bodies are two different environments, just like two different countries with different cultures, and therefore, although the general system may look similar, it cannot be treated the same. Just like bird species have different environments and different diets, people do too. Each has a unique life and should be treated as such, with nature assigning a manual and diet for the owner to look after.

We don't see lions stocking up on zebras or snacking on protein bars or cakes, nor do we see any other species. Allowing nature to do what it does perfectly and only taking organic essentials. Diet! Diet! Diet is the nature of the mental health epidemic, the inability to see life beyond the system, illnesses, infertility, disabilities, and more. We would not need as much energy supply to drive our energy prices sky high when unneeded offices and industrial manufacturing are reduced to minimums and necessities; there is no need for energy supply in large shops when we buy directly from local farms and focus on what our body needs. Just as there is no need for mass chemical plants or overproduction of medicines that have been fed to the animals or disposed into the ground; there is no need for more than essential medications when there are few illnesses around, or no need for mental health facilities or meds when there is no mental health, or no need for weapons when there are no wars, there is no need to fund climate crises when the climate is levelled up by us making different choices and who is causing hem.

Choosing to fix our own primordial diet and the choices we make will fix everything else, returning us to the ability to see everything from different perspectives, just as intended by nature, and all that's amazing about our beautiful planet, humanity, and the species we are. As much as many of us like to think that our life in this body will last forever, it will not. Yet, we oversee choices that will affect the time and nature of our body, which, with its magnificent team of cellular intelligence, will allow us to have this life experience.

The original form of the soul is made of energy and therefore, by its nature, is free, thriving for freedom while inside living matter as a human body. Our original form makes us confused about who we are, and why we are here, or rebel against human form over the need to learn its individually unique nature and understand binding rules that are imposed by nature on all living matter, which are the principles behind the building blocks of life and not the building blocks of the system we made up.

Rebellious behaviour towards Natural Laws eventually leads to suffering and displays in the form of illnesses to push the embodiment and evolution of understanding forward. Putting the elements of soul freedom and the binding elements of the Natural Laws of living matter together into a more powerful revolutionary and resolution perspective. Especially wanting to build future pathways that will benefit humanity more, which is scary to embrace at first. On the other hand, the positives are no different between swapping something outdated for something with greater power that's controllable by you.

For example, instead of buying into consumerism and wasting your precious life force energy to fund the system and exhaustion of resources, try the opposite and see what happens. Did you know many people feel cold from poor circulation because foods are so wrong that body cells struggle to generate energy essentials at such a low rate to keep the body alive?

Do you think dolphins or any other species with incredible nature-given transformative senses and powers would still have an excess of those abilities if their nature's given diet and environment were jeopardized? We all know what will happens to this species, if we feed them with inadequate foods or keep them in wrong environment. Or, if they were raised consuming anything else, then what did nature intend? Isn't it already happening? Life in the oceans, birds, planets, ecosystems, and animals on land are declining due to other elements being added to their diet rather than primordial, changing their environment and ours into decline. Isn't it ironic how single people's choices in food, drink, and selective consumerism can determine the cascade effect of the collective, serving as a decision-maker on whether humanity on this planet will live or die? The Law of Compensation is the effect of a cause.

7 CHAPTER

NATURE AND HUMAN BODY

The human body is designed much like nature. And, like simplicity, including food compositions, it can be recognised as not far from its origin, simple combinations, and order. It is designed in such a way as to run complicated processes without constant bacterial wars in the form of inflammation and a clash between elements due to overload. For instance, when cells need particular elements, will it be easier to locate them within a simple composition or a complex stack of everything and anything, not far from looking for a needle in a pile of garbage? Everything in nature is a constant fluctuation balance, a little like stock markets or human behaviour, exists separately and together at the same time, yet forever changing. Everything we need have already been provided by nature. Tweaking and interfering are different actions and result in different outcomes. Tweaking supports growth from a natural perspective, like putting a supporting frame around the plant or a tree. Interfering, which seems useful for a short while when prolonged, evolves into an accumulation of dysfunctions because that is where the balance tips. No different to the accumulation of arsenic in the soil or ongoing medication in the human body over time. The accumulation of anything not intended by nature will tilt the balance, which is precisely what is happening with CO2.

Anything we innovate and apply inconsiderately to nature's ways and laws will plant the root of interference, which will have a more significant adverse effect in the end than at the beginning. The same principles apply to the human body. The Planet is not designed to deal with interference, especially the type that tips the balance.

When an accumulation of arsenic in the soil occurs due to regular input without the abilities or extremely slow abilities of elimination, even small amounts will be added up at a price of health over time, just like a person who consumes rice more often. Accumulation will be more significant in the person who consumes it daily than the person who has it monthly. Yet some genetic templates and their biomes designed and are designed to metabolise some levels of toxicity. It is humankind's need for innovation and boredom or perhaps challenging ourselves at innovations, adding complications to our lives. We are the cause, and we are the effect. Naturally, any irresponsible creation without thought ahead of benefit for the self-inner team of cells, all living and for consideration of others, will result in failure and a dead end.

Physical hunger only exists with a lack of understanding of Nature's narratives about the precision of diet. Did you know that healthy adult who regains their food match with the NatalDNA template do not require more than one meal or a maximum of two meals per day? Desire to eat or constant hunger happens due to the absence of the required nutritional composition that is Nature's match for that body. Does diet truly influence our state of mind, abilities, potential, physical and mental health? Understanding the laws of nature and its architectural design is indeed understanding the ways of health longevity interconnectedness and life itself. Manufactured foods mixed with the particles of single-use waste we willingly use, expired medications we dispose of and so on make their way onto our plates, offering no recognition from our cells. The nutritional composition of each food will have a different presence and conversion of its biochemistry not only when mixed with other foods but also with the inner body environment, once ingested.

Mass farming and consumption of such foods do nothing for our health, leading to malnutrition, internal exhaustion, and expensive health consequences. It's time to recognize and protect the extraordinary life that sustains us, aligning our actions with the principles of minimal impact.

In nature's eyes, each body is a different form of life, just like each organ or system inside the body or just like soil composition in geolocations and climates, yet all designed to work on different resources. Each is unique just like the factories we create, where each manufactures its own goods, and works with similar settings but different forms of materials, with one purpose only to facilitate its life form efficiently. What this means, just like us searching for the best possible resources that will produce the best possible outcome and last with the least possible use or expenditure and waste including low energy or fuel consumption cost to us - nature does too. We meant to help nature to prevent the exhaustion of resources for all life forms and maintain nature's architectural and alchemic balance.

After all in nature's eyes, your body is a form of life that is designed and integrated to facilitate not only a huge scope of mass of almost 40trillion cells which is more than people on this planet but also have abilities to channel those cells into higher faculties to contribute into simple or complex evolutionary process working with all life forms. For that reason, each body gets challenged to find the most effective and efficient form of resources for itself at minimal costs to nature and maximum output to itself and all life forms. Just like for different products different materials would be required. Naturally, poor quality material will have a shorter life span, and higher quality will provide a longer lifespan.

The definition of good and high-quality food in nature's eyes is very often not even near to what we think. This is usually due to our upbringing, lifestyle and cultural habits, the additives in it and human madling. I cannot count conversations I had on this matter, where people were convinced, they have a perfect diet. Well, we are also usually convinced nothing wrong with our car, until the MOT has failed. Once again, something and that goes into our bodies in the form of foods, drinks, drugs, creams; that we think is absolutely fine - is not fine from the perspective nature of our body and the way it was made to function for its best abilities. However, it's a "Free Will" when it comes to adding mileage bringing the expiration date forward. The nature of living matter was not designed to be driven by fear of losing life but by understanding its individual manual, identifying unique built and pathways to prevention.

As I was writing this part, I heard on the radio about families accusing the system of failing to help people with addiction, mental health, disabilities, and lack of care. Well, any system is as good as we build it and as lasting as we tolerate it. As the old saying goes: "Leaders must first be a great follower. One is never truly fit to be a leader or qualified to make decisions, unless experience all walks of life, walked the path, spent time and gained own practice." Just like in accountancy, the balance – all about precision.

If the abundance of high-quality essential and organic foods is way too expensive for normal families to afford, as opposed to manufactured and modified, then it is the failed system that is responsible for the supply, availability of good essentials and education on the subject. It is also community's responsibility to ask questions, support local businesses, practitioners and Organisations that are massive contributors to wellness.

It is not responsibility of Organisations such as NHS and many other Organisations that pick up the pieces and attempt to manage the "effect" of a cause.

Offering funding to Organisations that work on the "effect" of a cause without funding preventative methods, human wellness and focus on a "cause" at its core will not end well.

Preventative lies in education among children and families on such subjects highlighting a personal responsibility with clear support, solutions, and supply of essentials by the system to avoid negligence.

I'm not sure if I'm missing something but isn't it ironic that drugs are free with NHS, but organic foods, rehabilitation retreats, preventative methods, wellness centers, children wellness activities and other aspects of holistic wellness are not?

Perhaps, it is time to change the view. One can only go back to the drawing board to make it better.

We do not need to grow economy, we need to level it by supporting what is right for us and Nature. Organically grown foods availability and profitability cup or none profit on such foods, education and NatalDNA food match and pathways to safe fasting such an amazing way forward. Because the diet make's our brain brow connecting us to higher intelligence, consideration and the wisdom of life among all species. One will never be effective without the other. Second would be required elements and/or herbs to facilitate work for the unique needs of an individual NatalDNA template. Elements do not mean any supplements that you think may sound good but what corresponds to your unique template at a time, which is also changeable. Nature gave us ways to master it all, find consistency and help those who struggle. This is the only way to spiral anything up and reduce the speed of internal acceleration. Please see the images of the brick wall, as an analogy, the way the body is designed to build and the way it builds when we are not on board.

Besides compatibility with the NatalDNA template, food must be freshly prepared because that's when it is in its life form stage, meaning freshly obtained. Remember such foods will work at their best bringing regular fresh updates from local and global systems into the body with minimal waste of time, energy, or resources. Remember different foods create different environments within the body, where its form makes a huge difference to the outcome and how much life force energy or its reserves it will use to metabolize it. Just as too much rain or too much sun will have an impact on the quality of the soil. In nature, the highest quality material for that body building blocks are of highest purity when in compatibility between nature-produced elements in the form of food and NatalDNA template with the least complexity possible. It isn't about good food, but good food for your unique body. In other words, some foods that are nourishing for you, are poison for someone else and vice versa.

Why can it never be the end of dieting? Diet is a natural aspect of life allocated by nature for one reason only to nourish and not just with anything but with compatible bacterial profile and elements within the food profile for a specific life form populated with almost 40 trillion cells residing inside of you to work with your own NatalDNA template. To understand this, one must step away from assumptions that the world that does not belong to nature is sold to you as true and step into the fundamentals of life and see for yourself through the eyes of nature and how life was built. Every living creature on the Planet has been carefully designed to be nourished with elements of the original design and clean water for one purpose only - is to produce high quality life form. Which in return produces and transmits life force energy that is the source of life, connects to all life and fuels the force of life within all living matter. Evolving into potential one never experienced before. Without living through last the 60 years and exploring different angle of our existence from the perspective of Nature's Mastermind, we would never comprehend nor consider the abilities within us are possible on a far greater evolutionary scale.

Wealth doesn't lie within the system we currently have. Old must be removed to re-build a new. Such ways of Nature. There is a different system building itself in the shadow, with a far better future for all life forms. We must remain hopeful. Besides and hypothetically speaking, why settle for a system that represents an electric car when one can have its own electric rocket, right? What an electric rocket is, will come to one's understanding in time, when the mind is fully fledged. It isn't an actual tech but Nature's expression of an actual plan for us as a species. Nature must go through an evolution in many ways, often sustaining substantial losses only to develop what it needs and translate its understanding from the View acceptable to those species at that time. All we created is to reference "cause and effect" aimed to re-build better using nature's fundamentals primarily at a core and tread carefully, so we don't repeat the same mistakes.

In the eyes of nature, we are like children who played with all possible tech, toys, and games, drove fancy cars, achieved wealth and fame, and binged on all fake foods and drinks with great taste, yet still in limbo of figuring out, what is it about life meant to be so exhilarating. The ones who thought they did figure this out struggle to sustain such an addictive and natural state of joy chasing other ways to feel it again. Often becoming the base for human greed, which is often not about the wealth but about the exhilaration of energy it creates and how it makes us feel, when we strive for more, unconsciously becoming the balance tipper for others.

The pension age was always designed around fifty to move in line with the balance of time to free space for the younger generation to step in and experience life. The elderly would serve as wisdom keepers, supporters and guides of roots to unity and oneness. Nevertheless, nature's ways are always there and are available for those who seek answers, ready to embrace and show their real power.

Despite the complexity of the era, we live in, there's an inspiring glimmer of hope that manifests from our ability to learn, grow, and make more enlightened choices. We are in an age where humans can author multiple PhD's, yet often overlook the profound wisdom of Natural Laws, or struggle to navigate life with common sense might seem paradoxical given our intellectual capacity, but it's a hurdle we are capable of overcoming.

Environment and the nature of gene switch function are inseparable part of Human Design. What is NatalDNA food? Source of nourishment for life forms selectively designed by Nature at birth to work with your unique genetic template. The template, which also underwent nature assessment and survival scrutiny during nine months in the womb to achieve the best possible collection of chromosomes taken from both parents as the primary set, known as active genes and another collection as the secondary that are dormant and serve as a backup set. Which is not necessarily a perfect match as primary, but in nature's eyes will act as "will do" to save life. Which comes in, changes and swaps according to environment and circumstances created by food and lifestyle habits. Which gets replaced once the initial genes or set of genes within the chromosome has worn out.

There are a few analogies that come to mind, helping to grasp understanding. I like the analogy of car tyres. When something happens to our car tyre, we would replace it. At a time of desperate need, we would replace it with a "spare". Now imagine, the specification of the tyre you need is no longer available and you have to use a tyre of another make, which is not a perfect fit, wobbles a bit but gets you to places. That or you have to drive using the "spare" until the end. Both scenarios are not as good as the original. Prevention and preservation of the original by reducing the acceleration of tear and wear are still the better options.

Another analogy that may help is the role of the substitute teacher, which isn't the same as a permanent one because it doesn't know children. Substitute teachers wouldn't be aware of children's personality traits, such as their strengths, weaknesses, emotions, or sensitivity. Nevertheless, will do fine for the job temporarily. However, should different substitute teachers step in daily for the whole year, children would not be at ease, experiencing uncertainty and stress.

The point is the preservation of the original team with the NatalDNA template is always the best fit. As the body grows older, especially without correct foods, switches and replacements for substitute genes happen faster and more frequently as the speed of decline increases until there is no substitute for the substitute. This is usually when we notice our body reacting differently, which can happen at any point throughout a lifetime, young or old, all due to the speed of the decline and how worn out genetic template was passed on initially.

I was somehow lucky to experience such at first-hand, always knowing when my dad's or mum's sets came forward. The same goes for overloaded recycling mechanics until recycled material is no longer suitable for recycling and has no other option but to be put into the ground. The body does the same, with toxic waste known as "free radicals" often either stored in fat cells or floating around the body interfering with order within amino acids causing poor quality copying within gene replication and as a result of corrosion within templates. This means, removing the effect of the cause must always be the first choice.

Well, when the NatalDNA diet is located from birth, it's even better because the body is equipped in all aspects for that life form to function at its best and harvest energy potential and all evolve from there at its highest. Suddenly, taking a load off the organization's global, designed to work on the effect of a cause, and potentially can divert to work on a better cause.

Here we continue with Nature's tapestry of life. For efficiency, this applies to only organically grown foods, natural environments, and habitats, which allows those life forms to do their part for nature too, contributing to global microbial and nature's life force energy exchange that happens daily, and preferably locally.

Why this is important? Per natural design, all waste matter discharged from the human known as excreta or animal or any form of living matter meant to be disposed into the earth per design including corpse. So, the earth will read and collect data to balance its own elements every season as it does and create microbial upgrades locally and globally working with the planet's water and ecosystems.

What happens locally? Depending on the soil bacterial biome in that local area, in spring and then onwards it will bring different combinations of plants that are best to bring balance to all its local habitat. This doesn't apply just to humans, for nature all life forms are equally important just as for your body each cell is equally important and it's about how that unity among cellular life sustains bigger unity of life such as your body. And, how the accumulation of unity among human life and all life forms on this planet sustains each other. Next time, one should wonder about its importance and should understand that yes for sure its life form is important but think from the perspective of being a cell in one's body.

Using Nature's perspectives of our origin as intelligent species, we can focus on importance to support preordained wellness within ourselves and all life forms by educating ourselves and making right choices; by establishing what drives degradation of our unique abilities as species and use preventative methods to address it. Regardless of the origin of such being introduced in our lives as a consumer. We must choose the best solutions, whether it be foods, lifestyle, habits, mindfulness, goods and poor quality goods, technologies that stop us from using our brain or our thinking from a peripheral perspective and remove human interactions or valuable jobs. Because only an approach of such nature will stop humanity from developing backwords as species.

8 CHAPTER

ENERGY AND THE POWER OF TRANSFORMATION

We possess some knowledge of the human body, yet do not fully understand the Nature of its engineering, alchemy, function, and purpose in the world of nature. The purpose of life, from the perspective of Nature, interestingly, is not much different to ours minus greed for profit. Just like us always strive to find ways of generating or harvesting energy with the least possible resources, least possible time spent as working energy, and least possible human resources involved, with the best roadmaps bypassing outdated and unnecessary red tape or self-inflicted manmade roadblocks and produce least possible waste in terms of human resources or raw materials. Nature does too. Nature's aim is the same, only on the cellular level, aimed at humans to achieve it to facilitate all life forms. Removing the biggest block is our old-fashioned economic system, which seems to no longer fit into the objectives of nature as an evolutionary form of life. It is Life Force Energy that fuels life. When it's lacking, it often contributes to the incompletion, exhaustion, and expiration of such a life cycle or struggle for outbalance damages we create. However, when cultivated wisely, human potential can make the impossible possible. A basic example would be between the beginning of the internet and the development of Wi-Fi and 5G in terms of speed clarity bandwidth and so on. If such advancement is achieved within human cells, which are driven by precision in foods and a clean obstruction and toxicity-free environment, multiplying per 40 trillion and multiplying by the power of unity, give us an interesting display of human potential.

The further away our understanding and choices we make of Nature as an architect and an alchemist of life, the closer we are to self-inflicted illnesses, disabilities, self-destruction by ill minds and doom. If you are tired, unwell, have a health condition, anxiety, autoimmune, gut issues, or anything else relevant to health - this has everything to do with the food that goes in your mouth, it's that simple.

"Smart materials" are nature's NatalDNA foods that, when matched, have abilities to charge the metabolism with fewer inevitable ageing processes close to zero. Slowing down the renewal by rechargeable energy is the aim, not speeding up by metabolising waste. This allows the elder person to be fully capable, even at an old age. While we chase righteousness and power, it is the quality of air we need.

Manufactured foods were presented as a form of innovation. Years later, we discovered addictive and unhealthy trades of such foods not only became destructive to human health but almost disable human abilities to connect and transmit life force cultivation of Divine energy to receive guidance. All due to the accumulation of waste and toxicity from within cells. We can clean our body all day long. It is the cellular pathways that are on low transmission. If compared to G, networks would be at the speed of the internet when it was first developed as opposed to the 4G network.

Prayers, affirmations, mantras, singing, the art of dance or poetry and any art of human self-expression from the heart, the art of creation with consideration and coherence towards all life forms requires clean energy fields as pathways to energy cultivation and transmission. The bird species and life sounds from the forests, oceans or lands are a perfect example of such.

Nature's way of reciprocation and balance requires higher vibrations, where speed and life force can only obtain such full power from inside out at the cellular level to be reciprocated from outside in. Giving and receiving. Just as water drops create oceans, our cells create our bodies and are cultivators and transmitters of energy internally and externally. Naturally, making each human storage and transmitter of life force energy to all life forms, receiving what one give's out. The very reason why, the nature of God's Divine creation in people always valued kindness, thoughtfulness and prayers as opposed to physical gifts. Nature built us in such a way that we all co-exist from a unified field, while a unified field is only achievable through coherence and balance of reciprocation.

Let's look at its mechanics. For instance, one can only be famous or admirable when others give their time, thoughts, love, emotions, and admiration putting them on a pedestal without realising that by doing so giving and sharing their own life force energy with that person, amplifying that person's energy field to attract more. Naturally, it is not the person who is famous but rather those who give their life force energy are the true powerhouse of the unified field allowing that person to hold that field. The moment attention is removed, the energy is cut off, just like no one turns up for the party, regardless of how fancy the venue is, the place becomes empty having no life force energy field attached to it. The balance tilts.

However, when one turns attention to the inner cellular intelligence team with its own cultivation abilities and reciprocates life force energy by sharing with nature, the whole planet and all living matter become one unified field, allowing you to attract more in line with all life forms, giving abilities to experience so much more.

Nature aims at the human body as a form of intelligence to become a natural pathway of balance among all life forms and precision of balance among all life forms. Both environments are inner (cellular) and outer (planet/nature).

For instance, the precision of required resources naturally in the form of food or elements has the potential to maximise energy production and quality within the cell allowing higher rest time without the need for food. Where reserves then can be used by cells to explore the potential of evolutionary processes and nature's interconnection and works as a species. Such as developments of extrasensory abilities and other pathways which originate from within the human body or other abilities that exist in nature and are expressed in other species.

After all each human body is another element of nature. Being complex element has higher potential among all species, which is rather wasted on an outdated system where many are mostly overworked, and suffer from mental and physical exhaustion, disabilities, ill or depressed. All is achievable through know-how, time allocation and practices. Precision in match will require minimal need for food distribution and balance with minimal resources and therefore minimal waste that adds to unnecessary cellular exhaustion, which will stop robbing and depriving the body of nutrients it needs. Such an approach will maximise cellular efficiency, and energy production at the cellular level and maximise any other means.

Hypothetically, I wonder, if on average cell generates 20 to 36 units of energy per cell but has a potential of 95units per cell with practice, multiplied by an average of 40 trillion cells, possibilities become promising not only to explore evolutionary perspective but reserves for the body abilities to direct its field to heal, repair itself or other means. Simply because the energy is no longer wasted on recycling and cleaning useless waste, we put in. As an effect removes the need for manufacturing foods/drinks that are no longer fit for purpose, clearing pathways for a cleaner environment and the need for resources.

Interestingly, most food modifications are tested externally, not internally. Without checking its influences on body PH yet, each human body will have a different inner environment that will react differently with the modified foods or simply have the "rock effect" on the body.

Each human body has its own internal environment, just as different NatalDNA templates, diet, and biome, because each is unique and built according to the nature of its design. Just as blue eyes, would never be identically blue, having other shades with ever-changing tapestry. We have no way of testing any food modifications inside each body's internal environment because each is nature's masterpiece and is different. Therefore, we do not know how the nature of that body will express itself and conditions show up in time. Just like a little crack on the windscreen of the car that we rush to repair, in human body can be a minor damage expressed in the frequency of minor symptoms, yet a mini crack left unrepaired, expressed in the form of illness or condition over time. While, somewhere in the process of a child being conceived, it will receive a reduced quality of genetic template, causing cultural and lifestyle habits to become hereditary conditions, showing up to 20 years earlier, wondering where it came from.

If one is to build something, the exact least of required materials would be obtained. Imagine parts, tools resources were missing, where one must go through a wasteland to find what's needed. Or order delivered something else often without any resemblance to what was needed? Immediately causing frustration, stress and distraction to the plans and building process, but more than anything tons of wasted energy that could be saved and used for other means otherwise. Imagine such display happening daily in almost every human body, driving cellular exhaustion. While cells have been forced to use something instead of its original requirements, which isn't as good and may not create as lasting an effect as initially planned.

Many similar examples are visible in the system we currently live in. Often lacking common sense or described in the expression: "Left-hand does not know what the right is doing", governed by either textbook or a "know it all" ego that loves the comfort of what is easier even if it is wasteful, damaging, does not serve the purpose - still will not budge. The principle here is the self-evolution of an individual action to upgrade its own system via the root of understanding the scheme of work involved, because by doing so outer environment will upgrade itself. Why? Because you no longer fund manufacturing of what harms your cellular powerhouse which then can be pointed to where and what you want to spend that power on.

The energy inside your body is just like the accumulation of wealth, once you start generating it, you can direct it to whatever development you want to achieve via the pathway of the body which becomes the transmitter of such power. The understanding wasn't there before which is required for administration of such resources to execute the pathway.

Just like building the rocket requires precision in each pathway of the project to achieve desirable results, the body's cellular powerhouse demands the same form of practice. Like in everything if one doesn't have the understanding and ability to read Nature's given map, the end results aren't visible. This precision is only possible in clean energy because waste is toxicity and adds weight and takes its potential backwards, the same applies to high toxicity from pollutants, chemicals, and disruptions to the genetic template replication order within the body. It wastes energy.

As it stands with the rise of social media many giving up their transformative power, without in-depth understanding allowing it to spiral down daily sending emotions all over the shop. The human body possesses transformative power hidden within cellular intelligence that works through nature yet to be fully discovered.

Just as we can transform through dedication and training into gymnasts, athletes, martial arts or develop other exceptional abilities, the human body is capable of so much more. When intercellular existence and intelligence are acknowledged as a team of conscious operatives and operated respectively in line with Natural Laws, interesting transformative abilities come to light. Just as neuroscientists' state: "Where the thought goes energy flows." In other words, pointing life force energy into cellular intelligence instead of annoying neighbours or online social, and thinking of such form as a tangible transformative force without limitations, it will generate a reflective attitude from within; and bring interesting events into play, one never thought possible. In other words, if you manage your thoughts to give much attention to each cell in your body encouraging the cultivation of higher life force energy, end potential that merged with the nature of its origin, becomes truly limitless. Especially, when the owner of that body masters the ability to function on nature's given diet and manual to cultivate its possibilities. Naturally, life force energy will start to flow easily, and its harvesting power will increase. Just as a free flow of water would generate a higher force of energy as opposed to the river polluted with junk. Each body's cellular intelligence represents the same water flow pathways within the human body. Once free from incompatible foods, toxicity and pollutants, the body will heal the thought patterns flooding with understanding, inner knowing, healing, opening, and expanding communication channels among cells and Nature.

There is a bit more to it though. Since the dawn of time, all life species had time for nature, simply because gadgets and destruction were not around in such quantities. Remember what is said: "Where thought goes, energy flows." Why does it matter? When humanity was not distracted by self-developed tech, as a species of nature we were giving back to her, by giving time. Spending time in Nature, watching the clouds, admiring her beauty, catching the scent of meadows, feeling the magic of winds and rain drops – all this is us sharing our life force energy with her, while she reciprocated back by giving us the acceleration of unexplainable joy and gratitude, just by being alive. Well, nature feels us, just as any loving mother would. Only wondering, where did she go wrong? Seeing all the children she raised and grandchildren she supported no longer even looked at her with admiration and acknowledged her presence, except for filling with litter her lands and interest in spending her resources wastefully, to monetise on. Even young children give more of their time to toys, gadgets and cakes than they do to her. Withering and forgotten, slowly turning into a wasteland, drowning in our complaints, in hope for the childlike fascination of her ever-changing beauty, enchanted seasons, oceans, raindrops, mountains, songs and whispering winds. Naturally, she feels disheartened about losing her life force energy from within due to a lack of reciprocation, knowing we don't even have time to care. Many of us that do are outnumbered to drive the change that she needs.

One can only master itself in achieving such flow, the rest will follow. It helps enormously to acknowledge the body as a borrowed form of intelligence that must be treated according to Nature's user manual. Learning to operate consciously and with much care. Imagine any thought pattern is an activation mode and remote control where accumulated life force energy is directed. Can you see the negligence and waste of life force energy on time-wasting thoughts, people, scrolling or events that are taking your life force energy away from you? And you are freely giving it away? Nothing wrong with caring and sharing life force energy. However, perhaps with the conscious allocation of such a powerful force? Naturally, many of you were not aware and that is ok. Now that you do, you may look at distractions in your life differently. Remember, time is the most precious commodity of living matter. No one can ever give you back time that's been wasted. Just as regrets are last minute and irreversible.

9 CHAPTER

NATURE AND PURPOSE

True exploration of the possibilities of one's insignificance into significance in the eyes of nature, is to harvest nature's gift by mastering inner tools and compass that will cultivate energy power with the least possible resources and the most efficient ways possible with the best effective ways and least cost to nature. I promise this will be worth your while. Once you achieve that, higher faculties will be open to you as a way of gratitude for listening. What I mean by that is the world with endless serendipities. The world that I live in and love so much; makes my day, every day, starting with childlike curiosity and wonder about how each day will show up and the magic it will bring.

Interesting observation is when children raised outdoors and with Nature the automatically attuned and often manifest many things at a very young age. Until body becomes polluted with wrong foods, believes, diet and lifestyle – blocking its pathways. All transformative power is in NatalDNA Diet. No wonder for century's teachings always stated "The body is the temple of the creation". Just like Mountain Rivers are created and transmitting The Life Force Energy when clean and unpolluted, our cells from within our body do too. Diet creates clean pathways for the energy to flow while lifestyle choices limits pollutants that effect clean water, quality of minerals and nutrients with our foods.

Allowing us to manifest is the nature's way to reciprocate what be adhere with respect to her ways and channel communication in all her glory. A place where wanting more will be growing inwards, mastering the next stage of human evolution and Nature guides us to bringing balance to the world that starts from within. Which then we must teach your children and others.

If or when you ever stack from not knowing your part in this world, remind yourself that you have a massive team of 40 trillion cells that are capable of wonders. Exploring possibilities by cultivating its enormous life force of power making you one very important life on this Planet. Just like each cell in your body, doing a very important part by helping others to understand this and discover the true force of evolution humanity is on its way to create. With the help and guidance of Mother Nature - the billion-year-old lady, the master of it all. After all, humans at the current speed of life can be looked at as a source of kinetic energy for nature. Nature will reward you with a fusion of vigorous life energy and a state of joy witnessing many miracles in your path that will make you wonder how this is even possible.

The moment one gains a true understanding of the code of conduct and power of Natural Laws, balance, and the clash, and dance of elements, and starts to wonder, ask questions and be excited to try; understanding its architectural design, its process, and its aim for end results, the energy of anticipation will become overwhelming with the feeling of being a kid again. Nature's pathway to true transformation is my favourite Law of "Cause and Effect".

While the contribution from all past human accomplishments certainly matters. Especially in the field of art, helping others to anchor, foster and cultivate life force energy fused with emotions of life. From nature's standpoint, our life's purpose isn't as grand as we've been led to believe, such as winning prestigious awards like the Nobel Prize or Oscars. Not in the eyes of Nature. Being able to weigh the pros and cons, considering nature's cellular viewpoint and from Nature's perspective is a valuable skill.

It is our energy that serves as a contribution to nature's life force energy that thrives to use the least possible resources to preserve life, which is for Nature is the most valuable commodity. To use everything most sparely and considerately and benefit from her power in return. Wouldn't be wonderful to think of something and have things, people, events, or pathways towards what you want somehow to appear. It is already happening for many who found their way intuitively but are not yet aware of how to anchor it and increase the precision they already have growing, expanding its abilities and benefits they can achieve by helping others on this path in return.

When smart template compatible foods create a positive life force energy spin within the body adding higher numbers to 86 billion neurons within the human brain increasing energy surge, none compatible create the opposite. One step forward, two steps back.

It's rather interesting how foods influence our biochemistry within our body being nature's living matter, which then influences energy surge, forcing elements to create more neurons (neuroscience). Which then, evolves into multidimensional abilities and growth in the field of consciousness for self, fusing energy into collective, followed by feeding it back to the cells in the reverse process. Ocean and raindrop cycles are another analogy, with rechargeable kinetic energy and vibrations created by the earth and oceans. All Nature thrives for life and we are in a quantity now making it possible to become a huge part of it.

The new model in human evolution of living matter and consciousness, which Nature thrives for and hopes to achieve is to restructure the system that will originate from each of us, at the end benefiting all life forms as a result. Fusing and driving evolution for all living matter, based on our findings over the last hundred years. Reflecting on lessons learnt, letting go of what does not serve, allowing to build a new. Creating anything based on Nature's fundamentals and freedom of choice responsibly is the nature of our true essence. Looking into the architectural model from the cellular perspective and the way nature builds our body and all life forms too, unconditionally. Showcasing through the tapestry of life allows our thoughts to wonder imagining what is truly possible.

Precision in food composition is the one that drives the energy cycle within the body into the world of living matter, living only essential, and returning to us from nature in the form of upgrades within elements in foods, creating even better form.

Everything we observe in energy production on earth is happening on the cellular level within the body and thrives for equally effective revolutionary advancements. Production of clean renewable energy that will store it within the energy body and use it as a rechargeable battery is one choice for moving forward. Another choice is to continue doing what we do raising levels of toxicity within the human body helping ageing processes and their acceleration spiral faster, developing further illnesses, infertility, and disabilities and killing our planet, which so unconditionally supports us. It's simple, man-modified, manufactured or simply non-NatalDNA foods drain the life force energy from within the human body.

Our body, when clean and attuned, works very efficiently, like sufficient wiring in a newly built house, and is preordained by design to guide us. It is only when your inner bio intelligence has been silenced with a drug, ignorance, or simply being unaware by the owner of the body that there is a manual to follow, that the body becomes exhausted to communicate.

Well, all we discuss in this book is rather logical from Nature's perspective thriving for precision in our daily diets. If we were to dilute a pint of milk with water, at which point would it be water with milk or milk with water? Sounds simple, yet a huge difference. The point is, at which point is our cellular capacity water with nutrients or toxins with water? When we were raised with wrong food habits, how would we know any difference? Only by obtaining our parents' illnesses 20 years earlier, which is the approximate timeline when the balance in accumulated toxicity shifts, until we pivot and start making changes. Or another time, that's when this book finds you! In which case, you are very, very lucky. I must say, once I discovered the power of my bio-team, I no longer understood the word lonely or felt alone. It was like I had a magical family inside of me that was entrusted to me, and I would protect it with everything I've got. Especially the magic it does for me daily and how happy it makes me feel. I get forever excited when people who come into my life want to give it a try because they get to understand and get filled with joy and light. After all, I know what awaits them.

My experiments and observations have shown what is possible with one meal per day, sometimes a maximum of two. I have taken time to build it, together with a lot of internal work. When my focus shifted to the internal cellular team, that daily giving me the chance at life, everything else became secondary.

Giving back by considering and caring for the health of my cellular team through the preservation of cellular health, energy, and nature's resources by reducing consumerism and overconsumption by understanding there many of us know nature has to provide for and catch up on already accumulated waste, is the least I can do. It has taken time to clean up my habitual ways and mindset to achieve accuracy and precision, but it is worth it. Remember this isn't done overnight. One must realise years of abuse went into our wants and needs.

However, it repairs to the best version of itself and becomes sharp, because it knows you listening and then your body starts guiding you almost magically. For me, as soon as I have anything that I shouldn't, almost within half an hour, my cellular team will tell me. When it happens, I experience a shift towards brain fog, tiredness, blurry vision, or low mood. The same happens not long after I consume either crisps, flavourings, additives, sugar, grains, or dairy. There are a few foods I'm very conscious of that I know are incompatible with my NatalDNA, including some vegetables, herbs, and spices. Its incompatibility with the individual body environment creates the opposite spin.

Did you know many genetic stamps have very few resources to convert dairy after the age of 2? Humans in general do not have the ability to convert dairy into life force energy. Instead, most dairy products and in quantities we allow, interfere with our ability to create clean fuel. So, removing it from the diet will be a huge gift to your super-duper intelligent cellular team.

Remember, it's not about you, but how well you look after your cells. It is all trillion cells working together to create a fusion of life force energy that lights you up with the joy of life for no apparent reason. It starts with you, and as you light up daily, others will shine with you wanting to achieve the same. Happiness is wonderfully contagious. Its efficiency of simultaneous action in unity and balance will produce that kind of power. To achieve such precision, use this book as guidance. If you struggle, sign up for workshops, and I will find a way to guide you until you are capable of continuing by yourself. Because you would recognize the precise moment when you aligned and connected to the nature of your cellular bio team that will guide you better than anyone, all the way to the top. The picture represents a relevant analogy to what dairy does to your body in the form of massive traffic within your system, along with the pollution that comes with it.

10 CHAPTER

NATURE'S BUILDING SITE

Nature builds the same way we do, or shall I say Nature expresses herself through us, pushing limits and guiding us, because we are Nature. When nature builds as an architect without interference, it will always thrive to build from the best template available based on compatible and available resources. That is where the longevity genetic template came from, which has become a rarity over time. However, if the genetic template passed on from parents became inherently fragile due to an incompatible diet from an early age. Then it can only be sustained with fasting and precision in diet to avoid the development of premature conditions.

Naturally, if a couple becomes parents later in life, the genetic template will inherit a weaker density template, as opposed to the firstborn, or having a child earlier in life. Some would say: "We were told it's ok to have children older." It is indeed ok, but it isn't the best from nature's perspective. One must go through clean up, nourishment and repair for at least 3-6 months to accommodate life both intend to create.

It also means that a child is advised to keep a clean NatalDNA diet where possible to prevent accelerated ageing and hereditary conditions showing earlier in life. We used to think that the nature of our body was so sophisticated to do everything which is indeed true, yet it was never designed to do so by doing its daily chores on an almost 2000% increase in added sugar, highly processed and modified grains, manufactured food, metals and chemicals in soil and foods, polluted air, and water and so on. Which is naturally common sense. Especially since the system we created bound by laws, policies, and rules. "Nature honours those who respectful to her laws". We apply the same rules and laws to the system we created. How can it then apply to one and not the other without the undesired effect? How can it be prioritised over the Nature of our very origin?

I often asked: Why some people can eat anything and have absolutely nothing wrong with them? Well, just like in nature, there are species designed to consume any variables some humans inherent to what I call platinum longevity genome. Never the less it does not mean it will not catch up with them. I've known a few that lived till their 90s but either in care home, with dementia or house bound. Having longevity is one element of life. Having longevity, mobility, freedom and a shine with joy from the heart, is another. The choice is ours.

So, never is subjective, because nature will always show its defects of genetic deformation one way or another earlier or later. It's just the way it is. We often assume we will have some sort of alarm ringing or some grand signs will show up to give us a grand warning. Sadly, the nature of the body talks via little things. We have been told to silence with a painkiller or ignorance, making somehow everything else more important than listening skills and the nature of the conversation with your body. Yet, the person may feel fine one day and become deceased the next.

So, as you may not see anything wrong with that person yet, it does not mean they are not struggling or will not experience health concerns later. Many are just tolerant of discomforts, always putting it down to age and keeping going. Reluctant to make any inconvenient changes. Ironically, it is never inconvenient to follow manufacturers' instructions on the type of fuel that must be used in the car. Knowing that is easy to get another car, yet impossible to get another body? Any influences of foods or drinks on genetic template create defects daily but escalates over time and when one is least aware, it shows up one day, like a burst in a water pipe.

Implants, such as hip or knee for instance, have become very popular over the last ten years, yet many are unaware that the body is a living matter, and it is constantly growing, while implant is not. Although looking after the body and keeping your own parts should always be the first choice, it isn't always an option when one time has passed, and the pain has become unbearable. Hip replacement for instance. Imagine when replacement happens, technically it's like changing a car tyre. However, five to ten years later, nothing has changed but the tyre no longer fits and feels too big for the car. How fast it will happen will be driven by the speed of accelerated ageing in that body which always depends on how well the NatalDNA diet is kept and supplementation to replenish what is lost. Remember where I talked about the natural law of compensation? The body takes deficiencies from other internal living matter that match in elements that it needs and compensates for by consuming or recycling its own tissue.

All is due to bacterial overgrowth that eats into the area because the owner of the body feeds it providing a favourable environment for it to thrive. As time goes by and a person doesn't address the diet and needed supplementation the bone density reduces faster causing misfit. This is because bone is a living matter, and if a person does not address its habits and deficiencies, bone loss happens at a much higher speed of accelerated ageing, as opposed to the person with a NatalDNA match. More so, if the person adds exercise without addressing the diet, it will happen even faster. Simply because exercise works on inducing trauma to replicate, while the body already struggling with the volume of replication material as it is.

While all forms of exercise have their benefits, when the NatalDNA diet isn't in place, either accelerated ageing is increased, or backup resources are used to compensate. It is just all about knowing. If someone is a coffee, sugar, or wine lover and those contribute to the removal of calcium, magnesium, and a few other minerals; then one should understand, that it must be replenished, or the body will take it off elsewhere. Inevitably accelerating biological ageing of that body brings the expiration date forward. Many rely on foods, forgetting that even when the diet is kept, the soil is depleted and our speed of life, radiation, and pollution exhaust higher levels of energy than the body can keep up with.

My personal experience of clinical death that came totally unexpected one morning after a cup of tea, kept me very aware of everything thereafter. Seeing people being well one moment and expired the next with completely perfect prior blood tests left me curious repeatedly. The same when it comes to cancer or any other illnesses. So, I looked for answers and when I found vague answers in the world of medicine, I turned to nature. What are the attributes of genetic density that contribute to the quality of the material built and the ability to last? Let's think of manufacturing, where the template for producing some car parts is made from precious metals like platinum. Such material has attributes of density, resilience, and resistance to interactions with other chemicals and oxidation in use, reducing wear and tear when templates or parts are made of it. But what contributes to its formation and strength?

Due to hydrothermal processes in the formation of platinum, from Nature's perspective, I look at reasons for hydrothermal processes within the human body and the reasons for high temperatures when we are ill. Higher quality proteins and Amino Acids of organic origin have higher heat resistance quality and stronger binding factors under hydrothermal processes, while nonbeneficial don't. These explain the health benefits of sunshine, traditional and infra-red saunas, hydrothermal treatments, and hot drinks when we are ill and raised heat and body temperature during exercise. All heat-related activities reduce bacterial overgrowth and toxic waste created by those bacteria. Toxic waste removal in the body boosts higher energy levels overall. Well, at any point when body temperature goes up due to illness or any other reason, the nature of the body's aims and function is to reduce heat-sensitive bacterial overgrowth, where heat, weakens and depletes growth of unfavourable bacterial proteins that interfere with replication.

Wrong foods and drinks cause bacterial overgrowth, which contributes to the mutation and reduction of higher quality genetic replication, and quality building material - the driving factor of many illnesses, body diseases and disorders including accelerated ageing and fast decline. Many causes of such overgrowth are driven by incompatible foods, toxic metals, chemicals, radiation and so on.

Interestingly, I recall, that my tolerance to the same alcoholic drink in a hot climate was better than in a cold. Naturally, the heat keeps bacterial overgrowth and therefore toxicity levels better under the heat. The metabolic rate is slower in the heat as well, so less energy is used. It will also explain why some people are always abnormally hot, the nature of the body trying to survive the recycling of toxicity and bacterial overload. Cold is beneficial too, especially contrast showers or cold baths.

Those who are always cold are on the wrong diet to them and as a result have difficulties in converting food into energy. Driving insulin resistance and elimination of toxic waste. Such often show up in form of a rush, allergies, urticaria and some skin conditions - all due to the wrong diet and body struggle to illuminate toxic waste. Being cold isn't due to poor circulation at all. If anyone asks me now. My answer is always the same: "The Diet is wrong for the body".

Disbalance within the system forces the body to dip into backup resources either similar to para-water molecules within cellular membranes to raise energy to generate heat that way or by use of adrenals. Either way, it contributes to the faster exhaustion of the body as functional system.

Just like exercise, without NatalDNA diet will push nature of the body to use fake bacterial proteins for tissue replacement building. How can it not, when higher-quality proteins for that body were not provided? Skin pigmentation and high levels of oestrogen seem to favour bacterial overgrowth too, hence sugar cravings. Potentially low vitamin D conversion, massive deficiency in essential minerals and vitamins group B's too. Let us not forget the creams many almost "bath" in daily which naturally create a barrier to the skin's ability to produce vitamin D. That's on top of other pollutants that block favourable life-supporting light waves. Besides, humans lived for centuries somehow without SPF?

My personal experiments and findings led me to stop the use of any creams, indefinitely. And, I used to be a skincare ambassador for 30 years, go figure. All I use now is a natural mix of liquid vitamins and elements that I mix myself fresh. Making my skin looking better now than all the creams, masks and serums I used for many years. Saying that, I also learnt effective skin mechanics which consume very little time and work like a charm.

Going back to the SPF. It is a personal choice but my findings indicate skin pigmentation or skin darkening - all originate from wrong diet, overconsumption of sugar or alcohol. I would always advise seeking shade between 11am and 2pm.

Let's look at which foods in general feed bacterial overgrowth or contribute to inadequate building blocks: wheat, modified grains, soya, all dairy including cheese, milkshakes and ice-cream. And, all good quality proteins that are mixed with honey, sweet sauce or added sugars.

While density is one very important contributor and the closest analogy to longevity gene, there is more, because nature never has just one magic pill as we would like. There are a few main contributors that when assembled make up a perfect picture. Some genetic templates can have core essential platinum but be challenged with minimal resistance to higher toxicity, the point where the body majorly slows down its metabolic rate taking into hibernation or the balance tips and allergies with other conditions show up.

For example, if a frame template is made of metal whose origin is easily affected by corrosion and it is exposed to the living matter and environment that are mass contributors to its structural decomposer and distortion, it will affect the end results of the required product that template trying to create. However, if material compatibility is favourable, the acidic environment that supports corrosive effects will be either no more or very minimal. Without causing any major defects in the process of further replication, providing higher sustainability without needing to waste life force energy on the living matter that does not serve its wellness. To conclude, the change in material to become compatible removes damages to the template to a very minimum. Simply because the body stops wasting its time and energy to recycle useless foods that are useless waste the body was never designed to work with.

Therefore, the quality of replication, template longevity and end results of the product would be maintained at their highest quality; this will allow the body to work on the expansion of life force energy from the sources removing fermentation to bare minimal, creating energy container from rechargeable battery, and slowing down accelerated ageing that causes faster wear and tear.

Why Does Expansion in Life Force Energy Matters?

What does it mean for the body to achieve an expansion in harvesting life force energy? Why does it matter? How does it relate to our food habits and faster abilities to manifest the life we all want? How does it contribute to the reduction and the need for a mass choice of resources providing fast environmental repair, health, longevity and last? This is something humanity has achieved in the motor industry, where the efficiency with which a fuel converts energy is referred to and known as "Fuel efficiency". Meaning: Hypothetically, if one car will use a full tank of fuel over 200 miles, the other will use a full tank over 500 miles. Whenever a fuel undergoes combustion, it converts the energy stored in it in the form of chemical energy to kinetic energy. This energy is used to perform any task. Naturally, when vehicles and parts are built in line with environment-resistant material, regular maintenance is in place and best-fit fuel is used, wear and tear factors will be down to minimal. The best part is that the example applies to the vehicle that is unable to heal itself, like the body does, naturally putting the body at an advantage when the precision of all needed factors is used. Who knows, we may develop abilities to fly? As humorous as it sounds, humanity has not tried to say it is not possible equally. It has the same tone as the conversation someone may have had about flying into space 200 years ago.

This is an example of the exact evolutionary advancements nature points to and guides us to evolve. When we stop wasting cells' energy on sustaining unneeded organisms that we feed, metabolising toxicity created by mass bacterial expansion created by wrong foods, that are hard for the nature of our body to metabolise, the body will start using that energy to expand self-healing capacity and other more meaningful and revolutionary advancements. Having an addiction isn't easy. Having an addiction to addictive additives that are hidden in foods is even harder.

However, having a cellular team that forever working on the repair and elimination of toxicity that forms faster than cells are capable of eliminating, is not a brainer to understand that either health conditions or the expiration date of that body will arrive faster. Even what we know as "mind over matter" is only possible when the body is in the position to build healthy cells for which building blocks are provided, as its nature is still a living matter, regardless of how strong the mind is. Precision in diet and narration of abilities to harvest higher levels of energy are further advancements nature of our body yet to be explored.

When we don't waste nature's precious resources externally or internally, such as cellular abuse, the nature of the body works with us at a higher power, reciprocating our understanding and mindful approach. I've come to understand how many unpleasant colds, health conditions like flu, gut problems, and other bundle of numerous health issues I could have prevented. If I had known their origins in advance were hidden in the simplicity of the Nature given diet. This would also have saved me time and money on countless visits to doctors, tests and chiropractors. Should I have known that my health was driven by every wrong food composition for my body genome in the diet I had. Simply, because health was preordained by Nature allocation within my diet that had been architected uniquely and inbuilt into each human design by the very creation, way before I came to this body.

Being in the world of wellness and natural longevity specialist, a nutritionist, skin specialist, holistic and spa therapies practitioner, science nerd, years of learning western and traditional ways along with hundreds of other hats. I'm too spent years in denial, when answers were in front of me, all the time. How did I not recognise it sooner? Despite being guided by nature from birth or despite always knowing, how do manufacturing processes and genetic templates work? Perhaps because everyone around me would always be confident that it was anything but not food.

I roll my eyes up, when I recall endless visits to a chiropractor, sometimes weekly, for spine manipulations, or sciatica, or having no answers on why my neck got locked, or both shoulders and all upper back were in permanent spasm. While now, since living on the NatalDNA diet, have not had a single episode of such kind again. Another example would be having facial treatments more than weekly, with the most expensive skincare and most advanced treatments on the aesthetic market, only to discover that my skin quality is much better now. Looking so much younger and flawless since I stopped using all creams and only use my autograph technique called "Skin4Life Method". My skin quality is so much better than it ever was and I don't pollute the planet with it! It's all about fundamentals.

Perhaps, like many of you, I was not ready to question, hear what must be heard, understand and see anything from the perspective of the Master of it all - Nature as living architect of life, relying purely on what I was told and sold. We all see the world differently and become awake to our worth and our own potential in our own time.

Since I came to understand the pathways of nature and architectural pathways, the way my mum's and dad's and ancestors NatalDNA templates waved into my body, knowing, that some hereditary conditions can be either pushed back for at least 20 years or may never show up, if the correct diet is in place. Seeing that many conditions that I had have left my body, having my body and my team of cellular intelligence guide me, seeing the results in myself and others; knowing I can push back health decline for my children as far as possible and help those who are interested in learning to take control of their health, is making me want to fly. Keeping to a diet is hard but losing people you love or seeing them ill is harder, especially knowing they don't have to suffer or could have more time. Yet, when you can see people that you guide get results it really is joyful and overwhelming at times.

What is a NatalDNA diet? Environment this composition creates and why it's important? What is the NatalDNA template?

To understand Nature's architectural and alchemical design we need to look at technical and mechanical details. I've attempted to simplify and explain the best pathways for you to understand this preordained manufacturing process that gets to be put in place by Nature, way before you were born. And how our food habits have a direct influence on how our body preserves and declines, quality health in the form of genetic template we pass onto our children, including the speed of growth and ageing. Below we explore variations and what happens when Nature's given architectural and alchemical plans for the original design are off track, disregarded or dismissed. Some examples and analogies used are hypothetical, from nature's perspective and for better understanding. This does not cover the karmic laws, but entanglement and transmutation waved among both, is something I cover in another book.

Nature of Accelerated Growth

The Natural Law of Time regulates the frequency of replication and the speed of growth and decline among all living matters. How does it work? The nature of accelerated growth is that it brings fast decline. This is especially visible in fast physical developments we usually appraise in children without realising it isn't something to be happy about. Something I'm to once highly thought of until observed otherwise. For instance, if a child starts developing faster than other children. Early developments that many parents regard as an amazing achievement will take the body faster to reach pick growth, after which the body will turn into decline.

We never look at any of it from nature's perspective on developments like the early ability to sit, walk, talk, first teeth appearance and change for second teeth, including adult-like maturity and early puberty. It also refers to the loss of a childlike outlook on life, including loss of playfulness and a state of constant happiness and joy. Why are fast developments in children in the eyes and reality of Nature nothing to be happy about? Why is it important? How is it relevant to health and the body's early decline outside of its natural time?

The Nature of Ageing

Let's have a look at the Nature of natural speed in line with Natural Laws and the opposite known as accelerated ageing. Everything in Nature has its time and its order working in line with all life forms. Naturally, there is a pick for growth before the body turns into decline. The only difference is that children who are at a faster growth reach their pick faster than those who are in line with its Natural timeline or a little slower. Inevitably, and as a result, they will move into a faster decline. Of course, it is wonderful when you see your child exceed in their studies until you learn why. Especially when you observe those parents who had to endure the bereavement of their child before them or become a carer for their child without even knowing why.

In Nature it represents the balance being out of sync with its natural flow, spiralling bringing the decline and expiration date forward. The Natural flow of life and balance gets out of sync when the NatalDNA template isn't compatible with the original material that's been preordained by nature to build, so it lasts longer. Why? Because it must work much harder, constantly cleaning up and repairing more than it normally should, squeezing energy even if it's meant for essentials, making ends meet without ever having any rest, fighting for life. In many cases, using up backup resources before its time.

If it's still not understood, take a good look at poor families whose lives would be a very good display of the state the body cells are in for many humans, while the food choices sack the life out of them.

One must remember and acknowledge that although your soul is immortal, your body is not and how fast the expiration date is approaching has everything to do with what you put in your mouth, on your skin, and the quality of the air you breathe.

Everything in the world has boundaries. Nature works on reciprocation. Although time is intangible, living matter can borrow its power via tangible actions that preserve natural resources, like fasting, honouring natural laws, and leaving minimal food prints, which is enough to alter or delay the cause of effects.

Simply because the cellular mass of a trillion cells is as powerful as the mass of a trillion people, which is as powerful as Celestial movements where an individual choice on a large scale can alter the direction of the individual and the world. It is important to understand this and give yourself credit for the power that you, as an individual, have.

The Role of Speed in a Fast Decline

To receive or revive time, one must give time. What is accelerated ageing? Accelerated ageing is accelerated decline. When the body is in a fast replication process biologically, then they should be otherwise, because building blocks crash faster than they should and need to be replaced. The best analogy would be to compare the lasting effects of a sandcastle, a plastic one, or a wooden house and a concrete one. The quality and durability of building blocks are unsuitable and inadequate to the original design; hence, wear and tear would happen faster. In addition, the body will be functioning at a limited capacity because incompatible elements will bind to elements that the body may need and remove them from the body, leaving it short.

Imagine you building a house and someone constantly removing concrete or good quality bricks, leaving you to build on limited and potentially incompatible material at a lesser quality than being preordained by the original design needs.

When this happens, the body becomes older biologically than it should be otherwise. A bit like a 2-year-old car with seventy thousand mileage on the clock. It also means the body's expiration date is coming at a faster rate than it should in the eyes of nature and the health conditions that come with it often would arrive early.

Infertility has the same root cause. Especially since now, many start families older. Well, if a person in their 20s has a biological age of 30s and a person in their 30s has the biological clock of a 40-year-old without an inadequate diet to match their genetic template. Please explain to me why would Nature produce life in an unsuitable environment with the wrong building blocks? NatalDNA Diet is essential for a healthy life, but starting three months or, ideally, seven years before conception yields even better results.

We live in the era of global pollution and are conveniently trying to produce life at an older age. How does it make any sense to think that new life would get passed on a good quality genetic template while quality diet and nourishment within parents are often non-existent from birth? Well, nature would always attempt, even if it means that physically or mentally life would be functioning in limited capacity and resources. Such an outcome manifests itself in endless medical terms we attempt to pin it to, while the origin lies elsewh ere.

Ironic ally and humor ously, this is

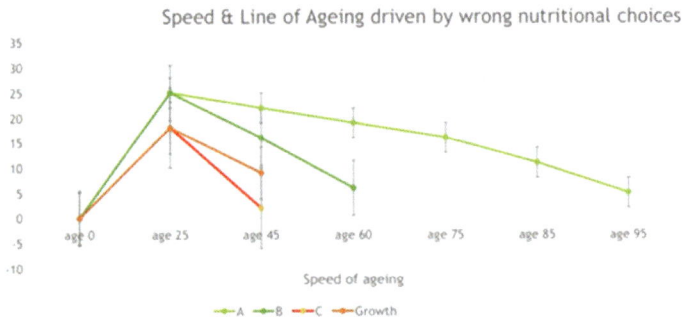

not different from the analogy when for decades we lived in either self-inflicted delusion or ignorance that the melting of the Atlantic Ice Cup has nothing to do with us, our behaviour, our consumerism, and global warming.

The assumption that something does not exist because it is happening elsewhere is as dumb as believing the body that is borrowed will leave forever without illnesses, even after it's been mistreated. Equally, consuming foods and drinks loaded with chemicals knowing it's harming the body yet somehow dreaming of a magic pill, telling yourself for years that a little bit of food field with poison is ok, convinced someone will save that body when nature calls it a day. Well, living in denial and without understanding, we are missing a much bigger picture.

People sometimes have the assumption that when it's time they will just instantly die without suffering. Well, one should be so lucky. The reality of the scenario at an old age and as it stands without being attached to a wheelchair, bag of medications, homebound or plugged into a kidney support machine in a nappy and potentially in a care home, is very rare. Bear in mind and I bet, every one of those people thought it would never happen to them. Ultimately, it is we who make daily choices and are responsible for lacking action and choosing to watch TV with take away, instead of preparing a fresh healthy meal. Inevitably, it is our children who would pay the price with early conditions and illnesses bestowed on them wondering, where they have gone wrong. Nature entrusted them to you to nourish and educate them on the laws of nature; so, they get a chance to experience a fully-fledged life without conditions.

The child trusts her parents to know what it must be and find the way to be educated, as with everything else. Nature's manual is simple, do not reproduce, if you not going to make it your business to learn the way of nature, essential building blocks and nourishment that the body needs while don't intend to look after the growing child that's been entrusted to you during this life. Sadly, feeding a child with ice cream for breakfast or sweets and cakes throughout the day, is not nature's way of nourishment, did you ever see these foods arise from nature or the nourishment of the soil?

Nature's Most Valuable Soup of Life

As oceans shape sand, sharpen rocks and facilitate everything we know as the soup of life, human plasma does the same within the human body. The nature of building blocks is to build, but not without a good bricklayer or without binding factors that hold it all. Blood is a carrier of oxygen and nutrients to the cells. Nutritional composition in the blood is made by the body from foods we eat daily, including plasma. The very reason we eat is to nourish. So, the body will use the food and attempt to convert it into building blocks for the different body cells it needs. Some compositions would be best for the skin cells, while different composition is needed for liver cells, hair, muscle, or gallbladder. The body will try to extract what it can and repair or build with what you deliver in the form of food or from backup reserves it has stored for emergencies. It also has a "state-of-the-art" recycling system, which is extremely effective but works best when a NatalDNA diet is used with fasting.

Blood tests are great and, like many things in medical science, can save lives. Nature is still ahead of us. The accuracy in terms of deficiencies and state of nourishment to slow down fast decline from nature's perspective is questionable. Did you know the blood test will show what is in the blood after the body has made what it needs and compensated from its internal resources like bone, muscle, bacteria or bacterial waste and other internal tissue? The body's cellular intelligence is constantly cleaning, making, and repairing something. General blood test results are after self-levelling and self-regulation events within the body; therefore, it will only show deficiencies when it is closer to exhaustion or inability to recycle, which is also a deficiency in converters or building blocks. The body works like our planet's climate in some way, as long as it has abilities to tab into something and convert into a substitute for what is needed to suppress whatever is brewing or irritating for the time being, even if it is a struggle, it will attempt to do so. Hence extreme weather events or what we know as natural disasters. As a result, blood tests often appear normal, although a person may feel unwell or due to expire. In nature is known as the tsunami effect.

As nature always operates in real time is accumulative effect of longer-term prior actions expresses itself, just as the flood of bacterial overgrowth that is driven by the inner environment becomes reactive, attacking the body before blood tests can detect it. The same sometimes happens localised or simply major replication template is broken beyond repair.

When precision in foods and its combination with the NatalDNA template is in order, this includes organic living environments, organic foods and no exposure to toxic metals, radiation, mould, chemicals, and any other triggers such as lack of sleep or stress which forces the body to use more resources; no one should be deficient in minerals or anything else. When minerals go in with the diet but are cancelled with sugar or incompatible food or drinks, the body will only be going around to compensate sending messages that it's hungry. In reality, everything that's been consumed in the meal has been discarded because of a few destructive elements that let's say killed essential conversion enzymes or removed half of the minerals that were needed to finish the breakdown and conversion for the other half of the meal. This shows how our food choices are unsuitable for our bodies exhaust body internal backup resources and technically waste the nutritional value of the foods we consume.

Take for instance organic bird or animal meat or organic non-modified grains, the moment it has honey, wheat, dairy or any other sweet base sauce on it, it becomes useless for the body in terms of nutrients. This is no different from trying to use organic compost for the soil mixed with processed foods and added plastic within it. Drinks like juices, fizzy drinks, flavoured water, and alcohol will have similar effects.

Ironically, we spend years in university studying accuracy and precision for surgeries, accountancy, architecture or aerodynamics. Yet, somehow think the same principals are not applicable in the world of nature and the nature of our diet or precision being matched with the body.

We would never dream of giving cake for dessert or a glass of wine to the lion with the meal. Right? Or feeding a cow with a chicken? Let's not forget the frequency of so-called chits. Our body is a species of nature and will remain such, regardless of what we tell ourselves or what we think.

How Does Uncontrolled Diet with Many Variables Affect Everything?

I like to take you back to school to revisit the chemistry class. Remember the simple rule in chemistry: some elements contribute and empower each other, others do the opposite, some neutralise and some masterful formulas can make things blow up. Each unique body has foods that work in perfect coherence and others are the drivers of inflammation and ill health. Remember that each vegetable, fruit, seed, or herb already has either a complex formula with everything you need in it or fewer properties but with the intensity of what you need in it. Let's also revisit the art of recipes.

What if one main ingredient was taken out or replaced with an incompatible one? Would it still be the same dish? The food truly is medicine, as long as the recipe has no ingredients that cancel the goodness of that recipe. Humanity is at this place now, where the body's cellular system is exhausted from too much stuff it cannot use. The rules are simple: once you achieve the state of precision in your NatalDNA diet, practice fasting, make sure foods are organic and stay with it, the nature of your body will show what is possible.

The larger the incompatibility of foods is, the larger bacterial overgrowth in the body environment would breed; the higher community of organisms your cells would have to support, means the harder healthy cells would have to work, the poorer quality building blocks would be and the lesser essential minerals your cells would be left with to use for them. Without intention of any discrimination, but for the purpose of example. It is like the system and societies where healthy working adults support the part of society who never worked in their life, draining the resources, and increasing the pension age for the healthy adults who do. This remark was not intended to incline anything but system defect and dis-balance at the core. Where if appropriate education, food habits and lifestyle balance were rectified and re-established within the system from before birth including education on responsible procreation, from nature's perspective reduction in numbers the system must support due to disabilities, mental and physical decline would decrease to a minimum. The nature of the body does not require proof or research paper to prove the obvious that is built within.

Microbiome is passed on to the child by parents, so within each body there always is a ratio of bacteria that can feed your body's strength and your body's weakness. Even after hydro-colonic treatments, the original bacterial pool would be restored within approximately two weeks. It would matter which one you would feed during the two weeks post-treatment.

We cannot be upset with our parents, system or old science as they did not know, but what we can do is make a different choice and pass on better quality templates and micro-biomes to our children along with healthy habits. We know unless bacteria facilitate conversions, too much of any bacteria will deplete minerals and nutrients because they need to eat too. Leaving the body short.

A larger population would require a larger quantity and quality of nutrients. Quantity of nutrients are those that are made by nature via local organic farming and which co-exist in adequate and shared balance with all.

Just like any pet, you may adopt big or small needs food, bacteria do too. As long as we don't feed unsuitable bacteria and the ratio of good is higher, the other will go into hibernation or do its small part where needed without overtaking. It will be kept under control by a compatible bacterial community and will do no harm. When fewer are part of a small community facilitating digestion and conversion, they never harm, unless the balance tips. Their presence, however, helps nature's global bacterial upgrades to know the community of the living matter on the rise or low to facilitate what is needed in keeping the balance. When our food habits take over the bacterial pool that handles natural upgrades gets diminished, which lowers your immune system or rather diminishes nutrients and resources that the immune system needs to stay on guard, such bacterial infestation party makes simple colds and flues into unbearable illnesses.

Reading your own body is amazing! For instance, I know now, that the moment I have frequent asparagus or blackcurrants, overload on dairy or nuts, within a few days I would get a weak bladder or cystitis. Or when I have gluten, dairy or a particular type of nuts, beans or legumes my eye vision, hearing would go worse. I can usually feel it happening a few hours after the meal. Or an Ice-cream can guarantee blocked sinuses or cold like symptoms, left shoulder pain or lacking cognition for me. I also noticed grains or legumes reduce my patience and focus and make me irritable, influencing my moods. It also had a huge impact on my knees, neck and joints, making years of dance school and 35 years of yoga uncomfortable and often painful. Thinking back, I always struggled with learning and focusing through morning lectures at the universities after having porridge for breakfast but got 98-100% on tests while taking those on an empty stomach. I was too nervous to eat, go figure. The Higher precision and simpler the diet is, the better and sharper the body communication gets. Just as a flower flourishes when the soil and environment well suited them, your body's beautiful self does too.

When bacterial overgrowth happens around Neurological or Neuro-muscular transmitters which in architectural design serves the same purpose as your electric wiring around the house. Water is the main transmitter of life force energy within all living matter. Bacterial soup overload minimizes or diminishes transmission function along with minerals and nutrients that facilitate the building of neurological pathways, leaving it short, merged or entangled.

Just like an electrician without enough resources had to save for essentials making the stretch. Imagine pushing the switch and instead of light music will start to play or the alarm will come on. This is exactly why having the wrong foods manifests in cognitive concerns, mind fog or pure memory. The restoration of the NatalDNA diet and fasting helps repair, restore and in many cases rebuild new pathways.

While we build one house to last for a long time, the body's internal environment is like the never-ending daily site of manufacturing, building, repairing, cleaning, recycling, delivering, communicating, policing, monitoring, collecting, and storing information and so on. Hold on, does it not remind you of the world we created and live in? Well, yes it does. The above does not mean some health issues will never happen. Especially since we have the last 80 years of footprints from rapid industrial growth. Nevertheless, it possibly be easier to treat and reduce the number of illnesses that can be controlled by diet.

11 CHAPTER

ENERGY TRANSMISSION AND PRIMORDIAL SOUP

Let's look at a simple example of nature's building blocks, order and influence of binding factors and the environment. Let's revisit the building story of the world's tallest building. Do you recall the origin of the environment that formed the sand? It needed the sand that was formed by water. Why is that? Sand is sand, right? Well, sand that formed from water known as primordial soup will have different shapes, quality formation and binding factors. In the nature of the human body, it is the human plasma that shapes the quality of the protein delivered or acquired by the cell. Each body is its own universe; therefore, it will have its own environment to do that. In other words, the same protein will have different end results or suitability in one body to another due to different environmental factors in each body. In those who are vegetarians, the body will be using proteins of vegetarian origin and recycling protein origin of internal microorganisms or another protein source within the body. Allocation of nourishing resources among all species prevents overconsumption and exhaustion of valuable resources, keeping the balance of growth cycles in place and is a win-win for nature and the nature of the body. Making energy transmission factors higher or lower according to that very soup.

Just like the ocean shapes the sand, it is a plasma in the blood that shapes the lasting quality of proteins enzymes and amino acids. Cells use those after it has been bathed in the plasma of primordial soup. And quality water shapes quality plasma. Each would have a different power fuse that comes from individual lifestyle and spiritual practice.

The quality of protein would be different from the quality protein of a bird, fish, fungi, rabbit, lamb, or grain. Besides, the body recycles internal proteins left from recycling microorganisms. The higher bacterial overgrowth, the lesser nutrients your cells get, because bacteria consume nutrients, leaving the bare minimum for the healthy cell to get by completing replication; In some way, such an environment will represent an undernourished mother-to-be; while at the same time, bacterial overgrowth doesn't just consume nutrients, it consumes water and takes life force energy from the healthy cell to continue to grow and expand because you continue to feed it; as a result, causing constant exhaustion, living behind toxic environment that disposed into human plasma and has corrosion-alike effect on the nearby tissue of other cells communities, going for locations with weaker genetic stamps in the body. This displays itself in the form of arthritis, podagra, neurological pathways, gut health and licking gut, most skin conditions, skin disorders, skin laxity, hair loss, bone or cartilage loss and the endless list of health issues. All taking us back to what we put in our mouths. Just like damage to the planet's climate cannot be reversed overnight, it takes around seven years for the body to fully replicate itself, replace, reinforce, and restructure its internal community.

Lectins, for instance, are suitable for those with the presence of a genetic template that is suitable to work with and resistant to toxins in foods that are high in lectins. For myself, grains drive insulin resistance because my body doesn't have what it needs to convert it. So, I learnt to rely on the absolute minimum source of protein, fast for my body to detox and recycle.

Human blood carries food and oxygen to the cells, it contains plasma which is known as primordial soup which contains many different elements known as "healing and building soldiers", and it includes and shapes the proteins. The quality of proteins just as the quality of the sand due to its origin will have different binding factors especially when the mixture of different origin of proteins has been consumed. Once mixed with recycled protein from organisms polished by inflamed plasma driving building blocks to mutate becoming more unstable and fragile. Giving birth to many other variants of proteins and amino acids. Small and slow mutations are natural, while fast-driven by modified building blocks create different effects. Well, as per Nature's Architectural design and logical approach in the way we build, if you are to build a castle with a mass variety of mixed-match building materials glued together, how strong, lasting, and resilient that castle would be?

The larger the variety is your diet, the harder your body must work, and the less time the cellular team has to clean up waste or rest. Just like keeping order over a smaller population is far more manageable than over a large. While the intercellular team has to manage a mass population that is 4.5 times larger than humans on our planet, we can see how precision in diet mass contribution to reducing accelerated ageing and overworking our inner cellular team matters far more than we think.

Functional defects come from wrong drinks/foods which dictate the environment ratio between ideal, favourable, or toxic.

Minerals get locked up in the soil the same way they get locked up in the body, this includes leaching; All causes exhaustion because once the waste in the form of useless food is in, the body has no choice but to push it through the system adding miles to the life cycle of genetic template just as it adds miles to the car. Or, just like planting seeds in the soil field with waste and expecting fully fledged fully nourished greens. Many health concerns in adults or children are not due to the absence of food, but rather due to the abundance of foods that create waste and make the body malnourished.

The longevity template will have high-density resistance to all possible environments. However, with few things in common such as no sugar including natural sugars, fasting, and slowing down in metabolic function with essential hours of sleep and naps. This is when essential groups of genetic templates that communicate with each other are still in substantial communication and living off the body mass, which explains body mass shrinkage that happens with old age. While the main genetic template has slower wear density and resistance to chemical toxicity with sustainability for its replication, that body will continue to replicate essentials, yet not without health conditions. Resistance to a toxic environment will sustain its life even if the owner of the body is on the bucket of medication. For other genetic templates, such a combination of foods and chemicals would be deadly. A little like different types of moulds, where some will be resistant to radiation and will serve as converters rather than sufferer from it. The same goes for when we build anything, we assess the environment, landscape, and material, carefully considering suitability among all required for the best end results.

When our approach to building anything is translated to the nature of building blocks and the way its architectural design has been preordained, it is easier to understand what's involved and navigate through narratives and decisions we make about foods.

Nature of juices. Remember it is all about Nature's allocation of foods and drinks for your specific template. If nature meant for us to drink juices, would we not have it available in the same way as water? Any juice within fruit or veg is meant to be consumed with fibre that it comes with or if its vegetables are in the form of soup, where its effects do not drive insulin resistance. In nature, having juice or energy drink is like having internal tsunami only on a cellular level creating internal sugar shock every time. No wonder the body becomes more obedient over time realising, its owner doesn't listen nor knowns its operation manual. A bit like driving a car without a driving license as long as it feels good. On occasion, juicing may serve well for some genetic templates that are fructose-dependent as trauma-induced boosters, yet regularity will drive wear and tear.

Microorganisms are just like people can change roles according to the environment they live in. Naturally, just like us, can be wonderful at supporting life when their nutrients dependent life and environment are in perfect balance, or the opposite, which manifests in taking over the system that pushes them to become destructive. So, it is the responsibility of the owner of the body learning about its unique form and take measures to supply with nutrients that help cellular team regulate the balance of life for themselves and others.

We need to stop thinking that we need to destroy harmful microorganisms and ask the question of why they behave in such a way and which environmental or nutritional deficiencies are driving them to change roles. The same goes for healthy cells converting to cancer cells, while cancer cells manifest disability of healthy cells or change their intended function. Perhaps, we need to stop thinking that an organism, virus or else are there to harm but the way nature communicates and displays functions and pathways that exist within the cellular network of the body and are meant to create a match with the external environment to facilitate life rather than attach it or keep putting it in the cage. And start thinking that all designed by nature has its place for a good reason, mostly to sustain upgrades and balance to facilitate all life. Perhaps viruses only attack an environment within the body that by nature considered defective or target the owner of the body to look after the cellular team better, value life experience in that body and to facilitate change. After all, we are still evolving as species and there is so much more that we do not know and therefore are unable to consider. Especially while us still at infant age with regard to the diet, considerations and overuse of resources. How can we pray asking nature to give us what we want while we are not reciprocating by doing our part?

The only thing that drives change in the behaviour of organisms just like us is the environment. The state of the environment is dictated and influenced by foods, quality water and air, chemicals, toxicity accumulation of waste including plastic, toxic metals, and radiation. The circle of life is simple and starts from the environment we create internally and externally because one will always mirror the other. In nature's eyes it's simple, making life or breaking down what's no longer fit for life, the very reason for microorganism's role switch.

Herbs for instance are helpers, facilitating environment for and in enzymic function, yet too can facilitate or cancel. When one is unwell, ask yourself a question about what environment you create for your cellular network causing microorganisms to switch roles between favourable and unfavourable. In nature, climate is the natural force of balance and facilitator of living environment for all living matter. Your body has its natural force too, but what drives the climate of your body being out of sync putting pressure on the environment? The answer becomes rather obvious. Overload of toxicity drives the balance tilt. This is also a major contributor to inaccuracy and inconsistency in our research and observation methods. Does expansion in Life Force Energy matter? Is it a key to ultimate health?

Diet and Mental Health

At the beginning, I spoke of young cellular clean well-nourished body which sets pathways for transmitting and reciprocating life force energy serving as manifesting line for many connections or doors in life. Yet, as person gets older the so called "magic" disappears. The truth is the body becomes polluted and starts storing stagnant energy, which then brings mental health, sadness, negativity, aggressiveness, stress and so on. It's like magic and happiness in reverse.

Well good news the source of joy and happiness doesn't disappear, it needs cleaning up! To allow life forces energy to flow. Fasting, NatalDNA Diet,

Massage, Meditations, Qigong or yoga are usually the best solution to address it at a core. As long as one does not think that it has nothing to do with food choices and does everything else but Diet. Everything else evolves from there. We often can use a short surge of life force energy by doing cold bath, running, keeping ourselves busy, spending days with friends, having alcohol, shopping or treats, feeling joy momentarily, only to fill that hole. Until we are back on that "low" again. Feeling everything and anything that is wrong with life, despite expensive gifts, perfect body, success, dream home or any motivational speeches and quotes – all short-lived; drowning in a constant battle between knowing how it feels to be joyful and high on life, but struggling to get that state all the time and keep it. All addictions come from it while one trying to feel the hole, without understanding the core origin of that feeling.

The truth is, and the answer is the same as the power of manifestation and clean cellular environment that makes us feel alive. This is how your cells expressing themselves from within. If you do not feel the joy of being "high on life", your body is polluted as it is not working on the correct fuel and is short on elements for your body. After all, everything we feel – is our body cells expressing itself collectively.

Years ago, I lived through all of it. Despite the fact that my life was and is dream come true and absolutely wonderful, I could not see it. Until I had the answers and persistently worked on it. I see the same in many people and a few of my dear friends.

Mental health happens, when energy connection to all life become stagnant. Once concept understood and the body rehabilitated with clean up, the person starts to feel life again, often better than before.

We become what we eat, drink, breathe and allow to feed our cells and thoughts. We also become the energy we cultivate through energy practice and feed our cells with freshly prepared and NatalDNA foods that are carriers of life force energy. When one prepares the food from fresh, if you can, use love and healing vibration frequencies music playing while you cook. Remember, your thoughts and your emotional energy go into that meal.

All living matter is like a sponge with the water and water is a mirror and reciprocator at the same time. Exact reason is "Grace" "Thanks" "Gratitude" before meal being part of many cultures for centuries. All are life force energy exchange.

Not believing in energy is like not believing in our own senses, love or phone getting charged. Nevertheless, one must have a good quality frame wear to transport, accumulate, expand or direct that energy where it needs to go. For the human body, high-quality frame wear comes from NatalDNA Diet.

Stagnant or life force. Just like the moon and the sun. Nature needs both. Dark brings up the light yet light, can burn when outbalanced or directed with focus into wrong source.

The environment is changeable, but some species would thrive in the toxic environment due to their genetically resilient built. Although, due to industrialisation and high pollution, genetic resilience numbers have been hugely decreased. Others will be unwell and will require a different approach. The same goes for food groups. As the environment changes, microbial behaviour changes two and with that behaviour cellular wellness, cellular system communication and the way the owner of that body will express itself. Anger and aggression, for instance, would translate into cellular rioting and struggle to cope with internal balance and deficiencies in essential building blocks of life to overcome such struggle and have a peaceful life.

I came across an interesting observation during the education and delivery of PRP and PRF treatments. A compatible diet will produce higher-quality plasma without any signs of inflammation, making wonderful building blocks for the treatment.

However, when the diet was inadequate, plasma quality was poor too. Making treatment practically useless. So, I will prescribe diet prior treatments and guess what? The quality of the material I need for the treatment will display itself, well enough for me to work with. It doesn't offer very lasting results; well, not as lasting as it could be should one wanted to know, but this is a different topic and discussion about aesthetics, skin, pigmentation, creams, hair and nails and all about beautifying body using Nature's alchemy that I cover in another book.

Naturally, many aesthetic treatments or repair processes within the body will have poor lasting results and sustainability, especially when the quality of the building blocks material is poor too. The body ages faster, mostly due to the wrong incompatible foods that go in it, accelerating cellular ageing. Well, most aesthetic treatments are based on creating a trauma to trigger a body rebuild and repair mechanism for which the body uses plasma and fibrin; naturally, effects would be short-lived without much lasting effect when building blocks serve as short patches instead of solid concrete. One should be very mindful of immediate before and after pictures, because one month, three months or six months later, results would be different unless the NatalDNA diet is kept.

Nature of Connection between The Body and Spirit

Spirit is a condensed form of energy and, just like life force, energy is infinite. The body is a form of living matter and is bound by The Laws of Nature. The fusion of both creates Life. Period. For understanding: Spirit in confidence form of energy can be represented by Nebulas or stars that we observe, while life force energy increased or decreased from repetitive movement like fish moving in the sea, waves currents or storms steering surging energy or freely flowed energy in cosmos before it condenses itself into denser form.

Everything is bound by the rules of Nature. The only thing beyond the rules is the ever-changing nature of flowing time encompassed by the Law of Cause and Effect. Everything in the world has boundaries, including cellular networks that manage the bodies we borrow or systems we create. Without it being chaotic, we cannot control nature because the body we borrowed is nature and is already bound by and ever will be in her control. Life and death are the highest boundary, and nothing can go beyond that rule.

The nature of embodiment or being truly spiritual means taking responsibility over the borrowed body, understanding the Natural Laws the body is bound by and the Law of the living matter. More so, it is not how long you will leave, but the quality of life you will have when parts of your template become irreparably damaged. Their part on soul origin, religion, spirituality, indigenous knowledge, and its embodiment discussed in this book will help to understand how those teachings have been guiding humanity over centuries with a weave into the NatalDNA diet for all species.

Without going off track, but for clarity and understanding of the glue between the body and the spirit known as embodiment. A short analogy of spiritual connection and embodiment must be explained. The best way is to imagine that from the time you were born, you were given a wild horse or another form of living matter to experience life, which represents the body, and your spirit represents the rider to become. As we grow, we slowly connect with the nature of that horse. Sometimes it masters us going wild, and sometimes we either master or learn to listen, trust, and understand its temperamental nature. While many of us look for a soulmate in a life partner or elsewhere, in nature, it represents embodiment. The bond between the soul and the body, where both connect on such a level to become one with itself and the world of nature, where one completes the other and vice versa. The perfect relationship between both the soul and the living matter, based on listening and understanding each other's needs, defines the strongest bond with which comes the experience of life in that body.

Those are lucky to use inner guidance to make adjustments, get to grow older and receive comfort, joy, and the longevity that comes with age. Naturally, because you love that animal, you will let it teach you its nature, its ways, and its diet, while you will share your abilities to create many journeys to share mutual experiences in return. If you don't behave forcefully in your own ideals, the relationships and journeys of life will always be short, painful, or difficult to endure. No one can be a master of nature. The spirit can only borrow its power while having this life. Full embodiment integration between the soul and the body on the cellular level, where both become one, usually begins between 40 and 50s years of age, where both finally become one. Some never get to achieve that, but that is a story for another time. However, it is worth mentioning that when the body is thought after and looked after as of great importance, the bond of embodiment is achieved a lot earlier, guiding one in all aspects of life gifted with joy.

12 CHAPTER

THE NATURE OF INTRICATE DESIGN

Health in all its forms, global or cellular, small system or large, is a representation of function at such definitive perfection that all life forms work at the precision defined in Natural Laws, on their own and with each other. My Nana once said: What you must know and keep with you throughout Life are the architectural rules within the Laws of Nature that are the core and govern all Life. Minuscular or grand, cellular, or universal - principles are the same. Precision is the ultimate balance. If you don't understand the meaning yet, just trust the process. By spending quiet time in your body, going inwards your heart, and connecting all cells within at sunrise and sunset, the guidance like tides will come as you trust her with the guidance acknowledging her knowing and respecting her ways. If it doesn't happen, it means your body cells are polluted, neglected, and struggling to transmit because they have that power—it was given with birth.

How many are aware that without Natural Laws, the power of inner knowing, and nature's guidance is blocked or opened just enough to stay alive, making them no different from the cattle stock? Not understanding why, one is following what they are following and why one is doing what one is doing from Nature's evolutionary perspective is a very grim place to be. Zest, or the joy of life, comes from this connection to all life forms and is given from within to those who dare try.

This means spending time to find foods your body is meant to work on and build with the best for your body and as per nature's design. If someone recommended this book, our educational programs with deeper understanding or through NatalDNA Diet programs, even better. As it will be from those who already walking the path of Nature's magic.

Just like rivers flow better and faster without mud or blocks, human body pathways are only clearer when nature-given adjustments are made to cultivate higher energy potential or manage its narratives, such as direction flow. It's what some recognise as "the power of knowing" or "going with your gut" by connecting to and from the Nature of your body's origin to guide you through all you must experience during this life. And, when you are not sure or in doubt, do adopt two lifelong friends who will never fail you to make it your inner "Cellular Team of Trillion cells" and a "Common Sense."

My Nana would often bring up the story of two trees; eventually, I knew by heart that its only understanding came with time. Many of you may remember it too. It was a story about the tall Elm tree and the small willow tree that grew alongside it. The Elm tree grew tall in the abundance of sunlight and everything available to it, focusing on the outside. Growing thousands of leaves and thriving to be as tall as it can, so in the wind, it will make such noise that it can be heard and admired from anywhere.

While the willow tree takes its time wondering, observing, joyfully experiencing time with another habitat, and feeling each moment of the life around it, it puts out long shoots to try to reach and enjoy the sunshine, but then drip down to create a shade for those in need. It takes its time, and just with enough nutrients, its roots are thin and can stretch into the depths of the earth for a better bond with Mother and to reach for hidden minerals and water. The willow will speak to the tall Elm sometimes. It would ask about the birds and the sky. The Elm tree would express itself admirably as an achiever and a conqueror of heights, and how wonderful it is to be a tall tree. Elm would suggest that Willow should focus not on its roots and its bonds but on the sky and how high it can grow to reach sunshine. "Well, my roots are deep and keep me connected with my Mother Earth whom I implicitly trust. She is guiding me and providing for me, as well as all life forms; she creates momentarily all life at once. While you are up there, she is looking after you down here. Reaching out to you too, through the roots", the willow tree said.

One night, the storm came. The wind rushed through the trees so powerfully that the Elm swayed to and from, struggling to sustain strength in its trunk, until it bent so far over that it could not get back up again. Its poor, weak roots could not bear it, could not outbalance the height to hold it, and broke. While the willow's roots were deep and held their balance well.

The storm died out, but the wind of nature came by willow once more whispering, "Roots are hidden, but they matter the most!" For many of us, roots represent family, which is also true. However, in this story, the roots are your body's nourishment origin, allocated to your body at birth, which requires precision and perfect balance internally to display its power and abilities externally in coherence with all life forms. The display of such compatibility originates internally, which is later expressed externally and defined as the inseparable compatibility of architectural alchemy by creation.

At first, I thought roots were a reference only to family, but then I was shown more. This was before I understood nature's alchemy and essence. Besides the values and family bonds, many hereditary conditions are rooted in the food choices hidden in cultural habits acquired from the parents, which express themselves in the same health conditions as a parent, later on in life. Without any thought, we label such conditions as hereditary, without a second thought, question, or doubt, that our parent's food choices may not be the best for them or us.

Naturally, they did not know then, but we do know now, and the root of a cause can be addressed. How many of us would find comfort in mum's cooking or childhood foods, especially when unwell? It is because childhood foods, not only provide the comfort of warm family memories but also the trust we put in parents to know what's best for us. Such lifestyle habits often become deep-rooted and hard to shift.

However, if one investigates, changes or tweak food composition, drink, supplement's that body needs, and lifestyle habits, the outcome in many cases becomes different without hereditary conditions manifesting themselves. So, in some ways, it's the quality and precision the roots are built with. The rest is an excuse and self-talk because it's our understanding and choices that fund, fuel, and drive the system we built, interest rates, investments, taxpayer resources, and others.

Our dietary and lifestyle choices, including addiction to consumerism, will create demand either to build an extensive number of unnecessary pharmaceutical plants with their endless chemical waste, expired medications, disability care homes, more hospitals, or nature's health retreats and rehabilitation spas with communities that field with love for the Earth and inner wellness. While doing our part in the world as well and focusing on raising healthy children by educating them on Natural Laws and spending time with them. Building the world where the "rat race," wars, and illnesses become the past. Naturally, if there are hardly any ill people, anything that has such origins will spiral out to a minimal need as an effect of supply and demand. In essence, NatalDNA health represents your body performing at its optimal capacity, a state that should be nurtured, sustained, and frequently upgraded to maintain minimal wear and tear.

Nature's pathways to health were designed remarkably thoughtfully and skilfully. Providing us with the liberty and capability to co-create the life we aspire to live. Imagine experiencing a natural 'high on life' daily! Isn't that far more rewarding than feeling low, ill, or depressed? It's simple. If you feel anything but joyful, aligned with your natural high, and in sync with life, you need to change your foods, drinks, and habits. The same goes if you have any symptoms, health conditions, accumulated weight, skin, nails, or hair concerns. All these come from the cells, expressing themselves from within.

When the inner world of trillions of cells built on protective proteins as opposed to destructive and filled with unlimited energy potential is happy, excited about life, and thrives, one feels the same from inside every cell of nature's given body. When the body is ill, overloaded with destructive building blocks, the most beautiful day is filled with clouds. Yet, when we feel exceptionally well, like in childhood, everything seems possible, even the ability to fly! Isn't it so? Children would usually easily feel joyful, playful and easy to dream of impossible, confident, and inspired by life because their body isn't fully polluted with junk.

The inner world of cells is excited to give that soul experience life, so it will push through all the sugar or manmade junk even if it must use backup resources that are meant for old age. Nature gave us pathways to achieve such precision without the need for substitutes, helping the body to remove overconsumption of unnecessary resources and reduce waste in return. This includes our habits and all we develop for self-indulgent self-entertainment, boredom, and profit that, in the end, find themselves in the ground, contributing to more waste that ends up on our plate and corroding our NatalDNA template. It isn't about having anything, but rather being selective of choices and mindful of quality that lasts.

When we throw away goods or clothes, we buy them cheaply and dispose of them effortlessly. Where do you think it, all goes? Charity shops are overloaded and can only sell so much because many have already accumulated the same. Imagine the planet was your house, where there were no bin collections but your garden. How much of the stuff that gets disposed of weekly would decompose, and how much would rot in the ground for a long time? Now understand and imagine if the planet were your house and the only way to dispose of anything would be volcanoes, and even that must be done without overdoing it. People just never used to have so much staff or such a high population. We often don't think of any of it this way.

Over twenty years ago, I was given a leather coat. Not that I did not love animals, but I had no understanding at the time. I still have the coat and I still wear it at times, because it is classy, always in fashion and I never get bored of it. A few months back I noticed someone gave me a look, judging me and maybe thinking that I was a bad person for wearing it. These got me thinking of how many disposable coats I would have had for the last twenty years that would be buried into the ground polluting our beautiful planet, its soil, and oceans, should I not have had that coat.

We often adopt views without peripheral perspective of cause and effect of other nature's architectural arrangements. Or our big picture is considerably smaller as opposed to a grander design. In my understanding, anything wasteful, is not far from a crime, simply by the rule of multiplication to how many of us on this planet there are. I also understand that our adaptation towards the NatalDNA diet may sound extreme, but is it really? When we only revert to the ways of nature and honour nature as an architect of life, taking responsibility for what we must do as a species, tweaking our health for the better to free up hospital places, and guiding our children towards better health habits. Is it too much to ask when resources are exhausted and human sanity is at stake? Perhaps, it's a starting point for those who are willing. Nothing in nature is done by force.

Choose experiences that create happy memories, raise your positive emotions, be joyful, give joy, and leave no waste. If you must have a piece of cake or chocolate, have it only after food, stick to "portion quarter" as a rule, be conscious of your body communication, and make it on rare occasions. Remember, many foods are appetite triggers, which means you feel like you need another meal soon after the last meal, experience constant hunger and need for snacks or feel cold. Feeling cold by the way, isn't a poor circulation, but an inadequate diet for the NatalDNA stamp of that body that unable food convert into energy. Thinking ahead, yet living in the joy of the moment, in line with the Natural Law of "cause and effect" and in line with the nature of our unique design, removes pathogenic behaviours and self-inflicted illnesses from us as a species.

Nature puts the same amount of energy, perfection, and effort into growing each grain, vegetable, fruit, or insect that contributes to the regular upgrading and building of our elemental nourishment or structure of our immune system. It does so by leaving behind local and global footprints each time its stewards fly or walk on the growth cycle of plants that we daily ingest. Just like the grain, bush, tree, or plant takes only what it needs from the soil to support its life, nothing less and nothing more, the same principle applies to the function of a healthy human cell. Some cells may require certain elements more than others and share their resources with others, however, overall working together in consideration and unity. They pass on what is not needed, convert, and pass on, sharing the surplus with other cells, microbes and life forms that live in body systems within. The opposite behaviours are pathogenic and represent the opposite of life. Cancer cells are switched behaviours of the healthy cell, usually defined by disability to produce a healthy functioning cell with the same noble properties and characteristics of conscious sharing.

As we spend years studying and understanding human behaviours, nature once again has it all figured out. Why some of us are different yet similar in many ways or why do we meet people that we relate to immediately recognise as our own tribe and others way out of it, especially when it comes to common sense that it's hard not to self-talk to calm yourself down from unavailability of such skills. Having X amount to agree on your point of view, while others being the complete opposite. Nature's architectural design as an analogy has an interesting answer. Imagine different people with their behavioural characteristics and views were to belong to one or another cellular team of a particular system or an organ in the body. Something we grown to relate to when we use astrological signs as similarities in characteristics. In nature, such traits are cellular teams. For instance, the pancreatic cellular team would have unique traits no other cells would have, the way they see the environment they operate in and what they feel is right for their system to be at its best performance will be hard to argue. All these would be different to brain cells, yet brain cells may not always recognise that forcing adrenals and other organs to work overtime may result in termination of life itself for the whole of the body including them. The same unique characteristics and equal importance would go to liver cells or blood cells.

The moral is in the eyes of nature, each human cell just as each human is equally unique in what role its life span came to play. Each outlook would be important because each would see life and the environment from the perspective of the cellular team it's from and therefore the views of rights and wrongs. The exception may be blood cells that seem to get to walk in all walks of life across the whole body. Gets to travel in every corner of the body to see where nutritional and repair struggles take place and reinforcements are needed and where not, where manufacturing is in defects asking for disaster and where it runs smoothly and in perfect shape. Immune cells will have their daily fitness challenges and collection of worldly data and knowledge on all-natural external and internal upgrades to keep all other habitats safe. At this point superiority and ego become the dust. Just as we can't imagine a liver cell having a conversation with a gut cell about its superiority, nature clears up for us that each life form is a bright star, equally important and has its role. Regardless of how big or small the role is, without experience, life would be non-existent or very limited. Learning from this example has taken me to understand that if our trillions of cells were to prioritize their economic, financial, or social status within the body because they ditched or had no understanding of natural laws, wisdom, essence, and alchemy of life, none of us would have a body to experience this life. This perspective is worth revisiting daily especially when our feet are no longer firmly planted on the ground and defined by ego-centric approach. We come to this life to experience, yet understanding why others are different will help to walk through life with compassion and ease.

On the other hand, cancer cells behave like weeds, consuming anything and everything in their path. We recognize this behaviour as harmful because it overtakes us, makes us ill and depletes the healthy cells or plants that we need. In nature, life, and death both have their purpose. Human life, like all other life forms, falls under "The Law of Free Will" to choose, allowing us the choice to treat it with negligence for short-term gains and as disposable, or to treat it consciously with wisdom, challenging ourselves to evolve in the way we never thought possible, learning pathways to expand our experiences within ourselves and with others. Just like those insects and drops of rain unconditionally do their part to support our currently worthless life. Human life is like a vehicle that can be used to experience the journey or driven recklessly to enjoy the drive from the Cliff.

For nature, each form, whether dead or alive, serves a purpose and is equally valuable. A dead body serves as fertilizer or food for insects and other microbial life forms, while a living body is a source of energy that contributes to the force of life to support nature. The choice is yours. In prayers lies the Nature of the Divine which will grant its supportive forces regardless, of whether the body is soon to expire, speeding to decline or not. The spirit is indeed infinite, but the body is a living matter and bound by the Laws of Nature that come with it.

The Story of the Banana Skin

Let's have a look at the ways we attempt to tackle the harm our daily choices make. Whether it is creams, vitamins, unhealthy habits, working hours, healthy foods for ourselves or the system we live in that creates the ripple effect for ourselves and the planet. Biodegradable or not equally harmful to Nature's alchemical design due to the volumes we create daily. Once again, we can twist and turn, but the answer lies in the precision of allocation. This isn't something that can be patched up or isn't something that requires proof, it's something that is defined by the Laws of Nature and is the way it is.

Let's imagine a banana skin being thrown into a meadow - the earth will convert it into compost within a few days and make use of it. But what will happen to the same meadow if we leave a truckload of banana skins? How long will it take to decompose? Will the soil composition be capable of restoring the original habitat? The moral here is: "Too much of anything is equally harmful and serves as a destructive force to all living matter, causing the life cycle to accelerate and bringing the expiration date forward to restart anew."

Now, I encourage you to think further to gain an understanding of the bigger picture, which is directly relevant to you. How much longer would it take for the soil to convert the same waste, mixed with non-compostable man-made materials such as plastic, chemicals, metals, and all cellular pollutants? And what impact would this have on the meadow's future? It doesn't have to be the meadow to serve as an example, it can be your garden or the park. The idea is to understand the principles behind it and how it is bound by Natural Laws regardless of the way we attempt to innovate. The Laws will remain the same.

When the earth doesn't have the tools to convert something, it merely leaves it there, waiting for better days. The human body adopts a similar approach to foods, including fats and oils that it cannot use. Much like reimbursing soil with materials it cannot utilise, the soil will continue to make do with what is available. This is because it has no use for something it cannot convert or lacks the tools to process. The less compatible the foods, the more the body uses its backup resources, which are meant to be reserved for old age. Just like a spare tyre or emergency generator is meant to be a temporary solution, not a permanent one. This is exactly where stress originates and where its destination lies — burnout.

When the body doesn't receive the correct building materials and the sleep it requires, it switches to emergency mode, panicking and creating internal stress, making the body restless, and work harder and unnecessarily. It usually expresses itself by making the person anxious and/or increasing appetite or having cravings, feeling like the head does not work struggling to connect thoughts. This stress then triggers an alert, allowing the adrenals to take over the body's function. Like an emergency generator or spare wheel.

When the required building blocks, including sleep, do not revert to their natural order, the body continues in this mode for days, weeks, and years, using its "spare tyre" or "emergency generator" depleting its own resources until almost nothing is left. This is our own fault because the system we live and work in left very little room for self-care, designed for us to push our limits at all costs since school. Then comes the crash, known as a mental and physical breakdown. All because the nature of the body was never designed to operate in emergency mode indefinitely. Just like a "spare tyre" or "emergency generator" designed to work for emergencies only.

In nature, too many ostensibly beneficial elements can massively tilt the balance and produce the opposite effect. An interesting analogy of this concept lies within nature's pathways. Even if something is compostable, other natural laws come into play. For instance, a single banana skin will decompose faster than a large bucket of skins. One banana skin will contribute just the right number of elements and goodness, subtly donating its properties to the soil, and merging with nature. Especially when disposed of in European soil since it isn't its natural habitat and the origin of growth.

The soil and bacterial collection of the soil have no familiarity with the foreign composition and in excess can affect the balance. For instance, the bucket of banana skins would be even more forceful, not only disturbing the balance of the soil but also dramatically altering its environment, affecting nearby habitats of other living plants, herbs, and beings.

Such an overload of something that in singularity and biodegradability and on occasion is ok, in larger volumes can be destructive beyond restoration—like the truckload of waste in the meadow. In this case, nature will scrap what it was and begin anew, composting in the environment that was forced upon it, improvising, yet wasting energy without need.

The point is that the same happens inside our bodies when we needlessly subject ourselves to foods that are not part of our nature, especially additives, flavourings, and everything under human food innovation. Companies with high profits need to answer to their investors and sustain high profits, always looking for ways to have us hooked.

Incompatible foods, herbs, supplements, or spices, either in original or cooked form depending on correspondence to the template would alter stomach PH and PH of PRP, enzymes, iron deficiency and quality of PRF and so on, altering the inner body environment to breed unfavourable bacteria which causes malnourishment, making the body work harder and cells deprived of essentials causing low immunity.

What does immunity mean in Nature's architectural design? Since the body builds itself daily, it only means if looked at from the perspective of a building, that it builds every aspect of the body system or organ without defects. The body would suffer when either alkaline or acidic because it must be in balance between the two. And, as we know foundation of balance is precision that regulates itself from the NatalDNA Diet. There are a few things that must be in place: fluids and nutrients in the diet which will influence your PH in the stomach and help to break down foods supporting digestive enzymes, iron and nutrient absorption conversion - changing the quality of building blocks and re-building compromised areas.

We live in a world where sugars are everywhere. I would advise removing all sugars, ice creams, sweets, juices, and even natural sugars, such as honey, and all fruit except berries. I would also remove all dairy, crisps, wheat, and legumes. Some NatalDNA templates are exceptions by nature but due to lifestyle chemicals and high toxicity, there no longer any exceptions.

Honey was indeed a fantastic remedy of the past but that was when sugar was not around or around in lesser quantities and when our bodies were in sync and in line with global nature's upgrades. Sadly, this is no longer the case. At least not until we repair the origin of a cause at core.

To foresee Nature's pathways and use them as preventative and reparative change is a gift that is given to those who follow Nature's alchemical origin, which I outline in Nature's architectural drafts and template selection. The change in choices we make outside our habitual wants and needs brings rebirth.

Rebirth brings an opportunity for you through your body to experience what is possible, a new life with a better way of living and the potential to create something magical that works in harmony with all life, just like a meadow did before it was ruined. Your body, Cellular Intelligence, the team of trillion, was indeed a fully functional and capable meadow at birth. This was before the times of industrialisation, an overload of chemicals and pollution driven by consumerism that led to cellular exploitation, dis-balance, and confusion. The trick is to retrieve, know your NatalDNA diet and master precision to maintain the state of the meadow.

The moral here is: "When one understands that every single element that enters a person's mouth, breezed in or applied onto a skin, nail or hair can be incompatible with one's own nature given perfection; creating a cascade of impact and will interference with environment state of precision and perfection in that meadow."

Nature has given humans an inner and profound understanding of how their bodies function. However, it seems we've turned a blind eye to these higher faculties, much like a bird burdened with stones that prevent it from flying.

Please do not forget that events such as the nuclear explosions in Hiroshima and Chornobyl, or any wars that have occurred in the last 100 years have had their contributions to everything that is on our plate, from the water we drink to the food we eat. Let's not forget mass fireworks and else we effortlessly dispose.

Nature does not care about our wealth, fame, political beliefs, or importance in this world. What truly matters to nature is how well we are attuned to her guidance, wants, and needs so we can treat the intrusted body according to its design; how wise, precise, minimalistic, helpful, and simple you are in your relationship with her – your true mother and guide during your visit to experience this life. After all, as you are experiencing your life through the wellness of your trillion cells, she experiences your life through you – subconsciously knowing your every feel, want and move.

There are other crucial aspects to understand. Within the realm of nature, everything is inherently accessible, granted, and consistently provided for us without conditions. Nature's intricate web of cells meticulously assembled your body, upon creation, —tiny warriors operating without cost—and it continues to be upheld without cost. The nature of your cellular intelligence team is to give you the experience of life in that body of yours with the bare minimum needs and the least possible cost. The principle of compensation adheres to the simple dynamic of supply and demand. Abundant supply leads to excessive waste and the depreciation of precious resources. Similarly, robust demand for anything gifted by nature, or that emerges from your garden with no harm inflicted upon nature, follows the same principle.

This elevated demand diminishes the necessity for significant financial resources, lessens financial interests, mitigates the requirement for stressful employment, and liberates time for cherished moments with family, the preparation of fresh family meals, spent time to breathe and the embrace of joy. This leads to a cascading effect of amplified sales and profits, as resources losing their demand are pruned from production, manufacturing, and the use of chemicals, thereby significantly curbing waste, and CO_2 emissions rapidly. All of this emanates from your choice to prioritise nature given diet and essence of life above all else and to adhere to the choices you were destined to make in this lifetime—contributing to the life force and resonating with exuberance.

Now, consider the law of compensation when high demand begets scarcity, affecting all forms of life, necessitating the reduction of species to safeguard existence and replenish resources. We find ourselves in an era where our preferences and disregard for resources, combined with detrimental desires, are harming nature, and generating considerable demand for precious assets like clean air, pristine water, fertile soil, and all other natural resources. This excessive demand, in turn, triggers a scarcity that threatens the well-being of all species, including humans. This is the outcome of the choices we've made.

The law of compensation, nature's unyielding mechanism, seeks to safeguard a select few species for the continuity of life—there's no requirement to protect every species. Merely a handful of pairs from each species will survive. Nature won't gloss over the repercussions of the harmful behaviours we've normalised; it will act decisively to safeguard life while it still can. As we relentlessly exploit the resources nature offers, retaining very little for its own nourishment, human bodies could serve as substitutes for replenishment and compensation, restoring what has been lost, thus enabling nature to commence anew.

13 CHAPTER

THE NATURAL LAW AT A GLANCE

For Nature, life, and death, both have their purpose. Human life like all other life for nature falls under the law of "Free Will" giving the choice of treating it with negligence for the short term and disposing of it or treating it with wisdom to challenge the ability to evolve prolong life and expand experience of it within itself and others. It is like a vehicle that can be used to experience the journey or run recklessly and be discarded. Each form dead or a life is for nature makes no difference and has its use. A dead body will be equally useful in the eyes of nature because it serves as a fertiliser or food for the insects or other microbial life forms and the body that is still living is a source of energy field that contributes to the life force of nature. Although, over recent years nature has been struggling with humans as a form of life escalating threat, distraction, infestation, stripping all possible resources and polluting all at a higher speed than the planet can restore or repair, becoming a threat to all life forms. We have become the pathogens, and the planet will use the same principle in function and approach as our body to eliminate the pathogens. Sadly, there is no magic pill besides converting our pathogenic behaviour into the behaviour of a healthy cell and in line with nature's ways, function, and wisdom.

Interesting, that regardless of uncountable life experience in all walks of life, books read, cultures and different religion's studied and explored, believe systems, astrology, astronomy, human design, science and medical science, kinesiology, acupuncture, yoga Thai-chi Qigong marshal arts or life force energy cultivation practices, holistic therapies, herbalism, reiki and crystal healing, sound therapies, cold bath therapies and more. I learnt one important thing. No one can be the master of Nature, nor can master Nature at its full capacity. We can only master ourselves and borrow its power through wisdom and understanding its essence to facilitate all life forms. Because, when something is gained, something is lost, or something is pushed, something is pulled. It can never work any other way or has been designed for personal gain. The Law of Cause and Effect always steps in to compensate: The principle of cause-and-effect highlights the interconnectedness of actions and consequences. It suggests that every action or decision we make has an impact, creating a ripple effect that extends beyond the initial act. When one takes life in vain, it will shorten its own or its family. Either longevity, health, or else, because nothing is left in nature without being compensated to sustain balance in one way or another. Karmic Laws of Life are like a shadow keeping it all in check. Everything taken or borrowed must be reciprocated through life force. This refers to extraction of resources, taking the life of a tree, an insect; or wasting an animal life in vain. Or saving the lives of those who would destroy the lives of innocent others. Nature regulates this through health and illness, wins and losses, achievements, and failures to facilitate balance.

In the intricate interconnection of life, we often encounter examples that reflect the interconnected nature of existence. Here are a few more examples in a similar context which we may know, yet not always understand:

1.   The cycle of seasons: The changing seasons—spring, summer, autumn, and winter—illustrate the perpetual cycle of growth, maturation, decline, and

rebirth. Each season has its unique characteristics and purpose, showcasing the harmonious interplay of nature's forces.

2.   The web and flow of tides: The rise and fall of ocean tides demonstrate a rhythmic pattern of movement. As the tide goes out, it leaves behind exposed shorelines, while as it flows in, it covers and fills those same areas. This exemplifies the constant exchange and balance between giving and receiving, as well as the cyclical nature of life.

3.   When one door closes, another one opens. This phrase illustrates that as one opportunity or chapter in life comes to an end, new possibilities and pathways emerge. It suggests that the closing of one door creates space for new beginnings. Old system spirals out new will rise in their place, including old beliefs driven through understanding. Doors are open through people, situations and millions of serendipities that appeared from nowhere yet serve as solid foundations in our lives.

4.   Yin and yang: This concept from Chinese philosophy can be unpacked in so many different ways. Represents the interconnectedness of opposing forces. Yin rotates clockwise, symbolises darkness in the form of rest, depth, in, close, slower pace, acidity, taking time, depth and expansion, femininity, passivity, and coolness. While yang rotates anticlockwise representing light, open, out, fusion, the point when the heart bit starts, fast growth in excess fast decline, alkalinity, masculinity, activity, a boost of life force energy to trigger fast repair within human body and warmth.

The two forces regardless of their names in different cultures are interdependent and complementary, and together they create balance and harmony. Interestingly and in the event you want to try, the pendulum swings freely clockwise if you ask to show you yin or earth energy; and anticlockwise, if you ask for yang or sky energy. In ocean gatherings, fish would swim clockwise when the earth's life force energy is low.

These few examples of many demonstrate how interconnectedness and balance are inherent in various aspects of life, reminding us of the delicate dance between opposing forces and the inherent harmony in our bodies, on this planet and in the universe.

Naturally, it's easier to cancel out and balance small amounts of waste and toxicity as opposed to large ones. Just like years of accumulated toxicity from sugar and food cheats harder to shift, as opposed to a few on occasion.

Using an approach such as less is more is the fundamental solution that contributes through sharing with all life forms. That is why not all the harvest is meant to be taken and stored. As per nature's ways, some to be left for that local soil and local species is to nourish its deficiencies and balance as it's not only us whom nature provides for.

When our body is unwell and suffers from higher levels of pathogens, what happens? Temperature goes up! Natural forces on this planet work in the same way. When pathogenic behaviour is sensed and suspected as a threat to all life, the temperatures on the Planet will continuously go up as the first call to action. Although we measure CO_2 for our own justifications for action on climate change and perhaps because the wisdom of the past somehow was lost, ignored, or misunderstood, it still does not mean "The Cause and Effect" will not take its toll. Whether Nature was discredited in translation into modern times or has no place in the eyes of science, the outcome will remain the same. It is all down to our choices.

The temperature on the planet does not go up for any other reason but to reduce the pathogen that is indeed - us. Because the human body is an extension of the earth, functions and behaviour are almost always identical, mirroring each other. Our million cells will mirror our environment on the body in the form of our health and our health on a mass level just like our cells will mirror the environment on this planet and vice versa effects: from inside out and from outside in - a reflection of all life.

Everything on this planet, just as everything inside the tech we develop, and everything in our body has its purpose and function. Gas and oil we extract from the Earth happen to have purpose and function too, serving as a shock absorber, heat deflector in the prevention of wildfires and buffer for Teutonic plate movements. Very similar to the functions of sebaceous glands, vitamin D, oils, and cranial fluid within the human body, and will have a destabilising effect regardless of our beliefs, if we do not act. Try driving a car without oil or fuel and see what will happen. Minerals and metals too that we are selfishly mine, have their own contribution to water cycles, minerals formation, conversion, and fertilisation. Stopping use on a grand scale, finding alternatives, stopping funding manufacturing of manmade foods, gadgets and garments is a big one and will allow Nature to start the repair process instead of fighting for its life and calling for the balancing forces of life. It is the demand we create as a buyer/consumer or has been sold as popular for demand - that funds all that is produced and contributes to the destruction of life.

14 CHAPTER

NATURE OF PREORDINATION AND

THE POWER OF BALANCE

The nature of the immune system is intricately designed not to function in isolation but rather as an integral component of the larger, interconnected exchange among the web of all life forms and the comprehensive global bio-system. As the integration between the human body and nature deepens, the reliance on external stimuli and interventions tends to decrease.

Sadly, chemicals have become part of our lives, whether it is in our water, in the air or food. We are funding by using them all the time in flavourings, additives, preservatives, soaps, detergents, fertilisers, anti-germs, on our skin, on our hair, we spray plants and animals, etc. So many chemicals and more are being manufactured and released every day that we have lost control altogether. Chemicals react with each other to produce other chemicals which, in turn, react with the old ones influencing bacterial growth that creates a new environment to produce newer ones! And all this without any of us adding anything new to the mix. Just like our behaviours change the quality of the air, soil, water, and foods, all we consume without thought changes the internal environment in the same way. All these chemicals add to the total load the body has to deal with and reduce its reserves to address other problems. Typical examples are the build-up of a resistance of bacteria to all the harsh anti-bacterial soaps, the increase in asthma epidemics and bronchitis, allergies, skin quality, pigmentation, allergies, liver damage and so on - leading to malabsorption and malnutrition which result in infertility, poor quality template reproduction and other illnesses.

In simple, inside each body, there are two pools of bacteria and it's all about whichever you feed and breed! One will facilitate life by helping cellular intelligence to produce life force energy, repair, waste removal, sustain nourishment and elemental nutritional balance for all life forms and work with your body environment; the other will rob your cells of nutrients focusing on itself, populating its own growth by consuming anything and everything it can without consideration to last, leaving toxic waste behind. These are the sources of malabsorption, malnutrition, and accelerated decline. This is when the environment within the body is out of balance and the body environment is rioting by attacking itself, manifesting itself in different forms of allergies.

Just like cells evolve from the bacterial pool you feed and breed, the nature of fruit, vegetables, birds, fish, animals, plants, and all species do, too. When the bacterial pool is dominated by chemicals, it is inevitably dominated by the harmful reactive oxygen species affecting the quality of enzymes and cells that the environment of the body will breed, which then will reflect its performance, life span and last.

The older population has a slower influence due to the nature of much slower metabolic rate and body regenerative cycles. It's worth mentioning that chemicals breed and dominate their own pool of bacterial population at a higher growth speed, contributing to faster decline and development of illnesses overall, especially in the younger generation. Making speed and slope steepness in the form of decline always changeable with nature-allocated foods yet always being driven by The Law of Choice (known as The Law of Liberty or The Law of Free Will)

Alignment is a perfect fit. The more harmoniously we align ourselves with natural pathways and NatalDNA foods as a reduction of chemicals, CO2 and waste in unnecessary natural resources, rhythms, and practises, the better equipped our bodies become to maintain health and well-being without excessive external assistance and mass reliance on pharmacological intervention that serves as growing barriers, but rather precision in foods, drinks, herbs, and plants. This does not mean we don't need medicine, but learning and working on pathways that reduce mass dependency on it by use of preventative and holistic means methods. Being educated and aware of traps.

We must not assume that events like the nuclear explosions in Hiroshima, Chornobyl, or any wars in the past century, which we have funded, did not contribute to the substances on our plates, from the water we drink to the food we eat. Nature doesn't concern itself with our wealth, fame, political beliefs, or status, but rather with our understanding and mastery of the natural and universal laws of life. It is through this understanding that we can navigate outside the systems we have created, guided by our hearts. Nature embodies simplicity and minimalism, like a nurturing mother who welcomes, provides for and guides us during our time on Earth. When that time comes to an end, she brings comfort and solace. She has given us a user manual for our bodies, which she constructed and bestowed upon us. It is our responsibility to care for and maintain this gift. We possess the freedom to acknowledge and harness its power, or to ignore it, but we must remember that Nature does not operate through force, and there are consequences when force is applied.

Nature, just like the car manufacturer that created your body, does not care what you do with it, because it is you who, if negligent, will suffer. Nature expects you to learn how to drive it, which fuel to use, know its needs and how to look after it. Once you have the body, it is your job to look after it and it will as long as you make it your business to know which foods must be used for fuel, herbs for health and frequency of ingestion or permanent daily fasting with one meal per day so it can serve you for a long time and without suffering. Each body, just like different car manufacturers, is installed with its unique NatalDNA features, such as template, biome, cell memories and diet in the form of fuel it must use.

Although there have been challenges, human intelligence has evolved, and we now possess a deeper understanding. It is not too late, and we have the opportunity to engage with nature in profound ways. Our actions should extend beyond recycling materials and economic growth; we must remove everything that harms the planet, all life forms, and humanity. It is little drops that create oceans, and it is little changes in people's choices that create massive changes in the world. This necessitates sharing resources, reducing waste, and decreasing our demand for anything that is not detrimental to ourselves, for we are interconnected with the environment, and the pathogenic effects we create have a ripple effect. To initiate change, we must start with our purchasing choices. We must remember that less is more, and even if packaging is recyclable, it still exceeds the planet's capacity. Earth will require decades to catch up with recycling all that we have already disposed of. To illustrate, imagine attempting to clean a house while objects continue to arrive faster than you can address them, store them, or utilise them. There is no space to accommodate them, and disposal becomes a challenge. This parallels the current situation.

For an experiment, one can try to store all disposable garbage in a backyard for three months, multiply by the population of the country you live in and revert to the story of banana skin. Where do we think it, all disappears? Considering that recycling is another form of industrial manufacturing that swallows resources of electricity and gas and still emits CO_2 via secondary pathways. It always makes me laugh when companies offer to collect recycling from your door, using cars with fairly high CO_2 emissions. Helping one way, taking away the other and often leaving more damage, is the paradox we live in. The same happens with elements and nutrients in the body when we consume the wrong composition. Interesting how our worlds are alike.

I often wonder why we fund technologies and their developments to help people, businesses, and societies have a better quality of life, yet it does the opposite. People seem to have to work more or the same and have less time and balance. Perhaps narratives are missing. Because, the functioning of the system, country, and society are set for the benefit of society, parks are busy not hospitals, children are happy daily without the need for a dose of sugar and adults don't go for the box of chocolate or rush to the pub to experience the joy of the day. The work is joyful and flexible, in line with the time we need to have balance without an endless race for life.

It is rather peculiar that we invest in technologies to be more productive, so we have more time, yet somehow, we work more to keep navigating those technologies and learning about new ones having even less time! How does it benefit wellness, may I ask? In an abundant world, everyone should be able to have the natural pace of life to enjoy. Having time to walk, adopt Qi Gong, Thai chi, yoga, reiki, breath work, meditation times, watch the clouds and stars, listen to birds, feel the wind, smell the rain or scent of the season and admire nature, have time to breathe and watch your children grow, study herbs, cook, help local farms, or grow your own greens, treat self to massage, reflexology, holistic therapies, and wellness. Explore the spirit of cultures. Spending life in shops will never achieve such an unmeasurable, natural, and lasting high. Having a more holistic approach to life is just so much more, besides the nature of the body will reciprocate in such a high and blissful state of joy that a shopping trip would never be comparable.

One crucial aspect is prioritising our health and adopting a NatalDNA diet. By reducing reliance on medications, processed foods, cosmetics, and artificial chemicals, we can minimise soil exhaustion in agriculture. Why would one need make-up to cover blemishes when blemishes will disappear via diet? Additionally, we must re-evaluate the need for food modifications and innovations that deviate from natural laws.

137

The waste and disposal of chemical and pharmaceutical manufacturing, as well as the practices associated with animal farming, also require attention. Even for those who require meat consumption due to their nature, there are alternative approaches. The growth of human abilities and extraordinary intelligence lies in embracing minimal necessities and adaptations. When consuming meat, if one must and it's required by the unique need of a genetic template, it should be done sparingly. Did you know that eating meat without required and suitable elemental composition is as good as not eating it at all? Because the valuable proteins in meat are wasted as soon as mixed with incompatible elements, animal life is wasted in vain, and the owner of the body doesn't get the quality nutrients either. All because we didn't know and usually until we are seriously ill, we don't want to know.

When life is taken even in nature has its rules and demands for balance. There is a huge difference when conscious food consumption is in place which is aligned with nature-given food range preventing overconsumption of animal meat and the need for taking life without the purpose of nourishment, and unconscious where life is taken due to ignorance of someone with a huge potential to excess depth of human intelligence. With this book, I will be offering free guidance for online communities for those who wish to know more and take back control over what nature originally created for them at birth to stay free. For instance, there are ways to achieve a nutritious meal with one chicken for entire months, instead of consuming the whole bird in one day and, in most cases, not even benefiting from nutrients due to incompatible elements.

Remember, in nature, even when a lion kills, many other species feeds on just that one animal. For instance, there is a massive difference when one family consumes 12 birds per year instead of 300 birds for those whose families eat chickens daily. The same goes for steak lovers or meat eaters, it is all about making sure that life is not wasted, and the animal dies in vain just because the chef in a new restaurant invented some Novo sauce, which unknowingly cancelled the body's ability to use nutrients inside the body environment. Forcing the body to dispose of the meal or put aside in the form of fat or stored toxicity. A few hours later, making the same person hungry again and again. Making the cycle continuous. Where the person has several meals per day while the body using own tissue to replicate instead of consumed food, leaving cells forever malnourished. Although humans are intelligent species to be able to recognise where they go wrong in daily lives but when it comes to health, we are own enemies, by accelerating self-inflicted decline and once ill wonder, where it all went wrong. Growing lab food will only detach us further away from the nature of our architectural design of the body. Nature did not design everyone to be vegetarian, and it is amazing when one can be.

When there isn't a choice and one must eat meat combined with fasting and hacking sustainability of your NatalDNA diet and health, results and nature of your body show up in a completely different light! Reducing the speed of body decline is one way to look at it. The volume of restorative functions and reparative processes within the body, where possible is another. Six months into a year into two years everything changes.

One starts to see everything in a different light, bringing joy, clarity of thought, understanding, and awareness of your body communication and a sense of achievement, purpose and empowerment, allowing darker shades of life to fade. Because it is no longer about you! You are giving back to the Divine, treating your nature's given body as intended, having a Global web of life and very Creation behind you, witnessing God's magic with endless serendipities, and saving lives too. The feeling is internal freedom with limitless possibilities within reach sharing this world, this life and this planet with others who walk beside us.

All is possible when balanced with wants, needs, and nutrients. This approach ensures nothing is overloaded, wasted within the body, or needlessly discarded via any means. It's about being human, producing the least waste, and walking through life with joy.

Achieving such a level of self-sophistication, higher intelligence, genetic dietary matches such as NatalDNA, and conscious evolution in the understanding of natural abilities represents a significant accomplishment, leading to the next level of mastery. It is important to recognise that it is we who fund our own illnesses. It is the man-made foods, unnatural farming, the waste we generate, lack of understanding and often ignorance disrupting our metabolism, fuelling endless appetite and in the end needs for drugs and its manufacturing, hospitals, disabilities equipment, care homes and adult nappies that come with it. Cells, which are products of nature, do not recognise complex foods that deviate from their natural composition, thus hindering communication among them. Our cells simply do not know what to do with what is not natural. It's that simple.

Ultimately, it is our individual responsibility to care for our bodies, which are extensions of the Earth and are designed to function by the simplicity of naturally allocated foods. Each vegetable on its own already has a nutritional complex without even adding anything to it. Everything else is our inability to understand or manage our boredom. Longevity should not be measured solely by the length of one's life but by the quality of health and overall well-being experienced. It encompasses the choice between a life confined to a care home, marked by disability and suffering, or an active engagement with life, finding joy in the simplest of pleasures such as the fresh air we breathe.

These outcomes are influenced by the choices we make through our free will. By refraining from consuming inappropriate foods based on our NatalDNA genetic template, we can remove emotional exhaustion, mental health issues, frustration, depression, and hunger, as all chemical reactions occur and originate from within.

More so, all areas of our life have the same origin which we know as health, but do not fully understand how it works. When achieved, it manifests itself from such a unique origin that natural change displayed by the intricate touch of forces of life will leave you with fascination, curiosity, and wonder. How on earth did you not know and how easily such a gift was handed to you on a plate? Only this time, without cost to nature, yourself, and all life forms.

Precision in nourishment changes the inner environment and the environment changes cellular health, which drives inner intelligence, life force energy and all the wealth and abundance one needs. By connecting to most sacred emotions, one becomes aligned with nature-given talents, driving the spirituality form of "plug-in" with the divine. Which then syncs into alignment with the timeline of natural laws that activate and drive passion, purpose, visibility, and mindset. Removes procrastination from cellular levels, creating and attracting clean and highly respective pathways towards relationships, places, countries and so on. Displaying nature's preordained pathways to ultimate abundance for those who seek.

15 CHAPTER

THE ALCHEMY OF BALANCE

For instance, from Nature's perspective, any illness, vomit, diarrhoea, or general gut issues are often representations of the clash among elements that indeed display a system being thrown out of sync. For instance, when incompatible foods and drinks alter the environment of pre-designed stomach PH or cause the lack of stomach acid or both, the effect of a change in the internal body environment further on will display malfunction by default. Correct compatibility and food composition will retain the required PH and self-regulated flow in the stomach acid without overflow or heartburn, repairing its process by becoming the opposite. When precision of elements is a must for healthy function, the opposite contributes to unhealthy. Just as in any system, it's just the way it is.

Apple cider, for instance, is edible and in the past was diluted with water, mixed with herbs according to unique species requirement and was used to treat plants that have been infected with warms. While now many use pesticides which are chemicals and unnatural, often killing the plant. Humans are nature and, just like plants, need their own natural herbs, spices or tweaks that will do the magic for their unique origin. Traditional methods are mind-blowing, but each unique and requires mastery, precision, time and patience. The same goes for breath work to facilitate expansion in lung capacity, cellular building site within lungs and cellular quality the body builds.

There population of bacteria, viruses and fungi living in our body as well as in our body. Let's not forget parasites. I had parasites when I was about 7 years old. Experience was traumatic, but I can see now, how it was self-inflicted with diet which led to poor immune system. Every life form requires its own environment and by nature the human body is not it, unless we create one with foods we eat. Even parasites are nature and have inbuilt senses to know a favourable environment for them. It is we who create such environment with our diet and lifestyle.

The cause of bacterial overgrowth would be due to the wrong diet for the unique bacterial balance of that body and an indigested or so-called "rotten food" effect. It would often happen when something as simple as stomach acid is out of sync due to additives, sugar, and incompatible foods, removing or disabling the bioavailability of digestive enzymes, making even nourishing meals become a waste. All because few elements in the diet cancelled the goodness of other elements, or the diet was completely wrong for that genetic template. For instance, cranberry sauce is meant to be plain and simple, mashed up cranberries with citric acid or lemon juice aiding digestion, especially proteins. Growing up, I recall cranberries were known as winter super-foods. Cranberries contain a large amount of vitamin C, so cranberries are widely used to prevent colds, and vitamin deficiencies and, in therapeutic nutrition, help improve immunity and are very useful during the cold seasons. When summer berries are out of season, cranberries step in to aid. Like all wild berries, wild cranberries are considered much healthier than farmed or garden ones. Remember nature's tapestry, and the bacterial and insect contribution that goes into it, driving the strength of our immune system. The large amount of benzoic acid in cranberries allows them to be stored for a long time without processing. Naturally, without adding anything to it, it is already "the miracle berry".

Wild cranberries will grow in swampy, wet, and mossy areas, far from the city and industrial production. We had to watch out for the snakes and their nests when indulging in cranberry picking. Wild cranberries grow in wild bacterial diversity as well as in an acidic environment with PH 3.5-5.5 making their contribution to winter health highly valuable. Not without moderation, pure form, and other elements within the diet.

Many use slices of lemon in tea assuming it adds vitamin C to their elements without realising that vitamin C is highly sensitive to heat and in hot tea will diminish. However, when tea cools down, a slice of lemon and its elements become very useful. Honey should not be used in high heat too.

By nature, each body has its own pre-set levels of healthy PH in the mouth, stomach and so on to correspond with nature's allocated diet. Stomach acid is designed to be highly acidic to avoid the "rotten food" effect. PH in the mouth as well as in the stomach has a sort of disinfection effect on the dirt or bacteria that goes into the mouth. Reason healthy tongue from tip to top must be always pink.

My Nana lived on a large farm since 1917 and raised thirteen children without any antibacterial soaps, sprays or else. Let's not forget children always put everything in their mouths growing up, especially toddlers. Adult immunity is built via years of growing up collecting and registering seasonal data from herbs, plants, fruits and vegetables in local habitat. Alcohol served as an anti-bacterial where needed. The nature of the body was designed in such a way as to familiarise itself with the world. Nature's given stomach PH for analogy can be compared to 100% of the surgical spirit effect. Correct PH for that body's stomach or else would be only sustained with the correct diet and elemental composition within that body.

When foods are incompatible or manufactured body does not know how to read its fake guidance, therefore how much the acid to produce puts the whole digestive system out of sync. These per Nature's designed manual would lead to changes in gut-friendly bacteria, malabsorption, malnutrition, and constant hunger. When the initial cause is addressed, the cascade effect would facilitate its repair, and a person can easily live on one to two meals per day, being perfectly happy, and enjoying fasting without ever-lasting hunger. Without causing over-exhaustion of the body's natural resources, accumulating upper reserves of energy drives the state of natural joy. Hope you can see now how a few wrong foods that are incompatible with one's body can alter PH, causing a cascade of other defects. Going by nature given a manual, when such continuous can only result in an acceleration of systems wear and tear, illnesses, disabilities, mental health, and hormone imbalance and so on. I let you fill in the gaps. Everything in this book is explained from nature's perspective as an architect and alchemist of life only and is not designed to diagnose medical conditions.

What is Nature's alchemy of balance? Observing Nature, everything has its place to contribute, yet when she must strike for what is needed to defend the balance. Rain on its own is fine, and wind on its own is fine.

However, when the two are mixed while both are at their highest force, it results in a hurricane. Saying that, when the two are mixed while both in the lowest force, it results in a gentle, intricate touch with abilities to get along. This is a typical example of the system we created many years ago or so, which was working then and only while it was in a gentle flow of balance, only applicable at the time. As the system reaches its maximum force, it outgrows its application and is no longer fit for use or relevance. Just as hip replacement no longer will have that ideal fit ten years after. Especially if the body cannot replenish minerals and elements that cause the bone or cartilage loss, fast enough. Which caused an initial decline in the first place beforehand.

Another analogy would be the technology. Which is only useful for an essential and gentle presence in one's life. The moment it takes over or becomes overgrown, one is spending a lifetime managing all that tech increasing productivity for what? Giving your power, energy, and life to imaginary success? The fact you are a life means you are already a major success in the eyes of nature.

For instance, if you had to do five tasks and do these well have time to enjoy life, be joyful and be happy and have balance; or have ten apps that will allow you to complete a thousand tasks exhausting your life force energy; which would you choose?

Well, the NatalDNA diet isn't that different. Few natural vegetables with good fats and few natural proteins in their simple form and fasting are easy for the body to manage, leaving energy reserves for more. Yet, once loaded with endless twists of Novo food tech, the body is exhausted to even signal you the slightest sign of anything. Almost knowing you would ignore it anyway, a bit like your boss with a deadline on ten apps with a thousand tasks. Or like a person, in the olden days, was tied up and dragged behind the running house. At the end wouldn't dare make a sign. This is also a typical example of a collision among foods, medicines, metals, viruses, microbes, and chemicals within the body. Giving the power of life for what?

The simple and vital importance of the micro-system between NatalDNA food profile compatibility with the NatalDNA template, NatalDNA biome and the natural order that the natural design of each unique body is bound by is simply unmeasurable and priceless.

Since one does not exist without the other, the crush and collision are inevitable unless the system changes its course. Since Nature is a master and an architect of life and does not benefit from our economic system, which is only based on our own wants and needs, it has already been written off by nature as unfit for the planet's health, where human wants and needs have become pathogenic. Unless we collectively abandon the sinking ship, change our course and habits, and build from scratch, the crush and collision of the current system spiralling down is inevitable. Nor does Earth's Climate system want us here.

If we continue this path and attempt to cheat the climate, the future does not look bright either with the current system because there will be more disabled people with poor physical and mental health than those who are capable of caring for them, performing jobs, use tech, pay tax, satisfy profits and so on. The scenario is very familiar to many people. Is when the head works without the body and vice versa, but not both.

It's a lot like creating a rule, usually made by government to make disabled people work without ever experiencing and knowing what disability is. Yet, the growing economy and funding systems and products that are the cause of that disability being unable to tackle.

How does one can learn to understand a learning subject when it loses its ability to follow the thought or connect the dots? Such effect displays spiralling down with much fewer taxpayers than funds needed to sustain it, making it impossible to contain and spiralling out of control, crushing the current system, which will become one large hospital, mental and physical disability ground with global care homes serviced by who? It's exaggerated, of course, but not far from where it all is pointing towards if nature gives us the chance at surviving global warming. Since one does not exist without the other, one will not repair itself without the other, and since every life form does not exist without the other, we cannot have both; that time has passed.

It's not all doom and gloom, though. There are solutions and the change, of cause outlined in the laws of nature and expressed in common sense. The Alchemy of Life is ever-changing, with each few hundred years bringing reflections and adjustments that were lost need to be found. Discoveries that won prizes at first, with time, reveal themselves in a different, originally unknown light, leading to a wish to turn back the clock of time. This was the time when the modification of Nature's wheat and other foods has taken place. We feel so big above Nature when we innovate without consideration, yet so small, when she decides it's over and takes the body back. All is learning, but only if we learn from it and change our ways. We often get so excited to innovate without realising that we still have a long way to go to match the true Master of it all—Nature—as the architect of Life.

The environment we live in shapes our beliefs, which, in turn, shape our perceptions and influence the choices of actions we take, resulting in the environment we create—an ever-changing environment of "cause and effect." In other words, anything we discover today or up to date and consider a scientific fact will be overwritten in time with a different fact, due to something else. Simply because as we change the way we observe and see, the world changes. It is us and the environment we create that facilitate and influence that never-ending cycle of Nature's Alchemy of Life and change.

We are particles and participants of the life force on this planet. It is we who accelerating our own decline, and it is we who can change its polarity and align itself with nature and the natural flow of life, allowing anything else to diminish itself and allowing it to stay what it must.

If we spent as much time and energy educating ourselves on Natural Laws, the basics of Nature's alchemy and what is important as much as we do on our own drama, in no time we could pivot towards the unseen force of life, we never thought possible.

The same principle applies to the ever-changing environment within the human body, which is driven by the foods one consumes every day. These foods generate low or high energy that projects into the body's environment in the form of one's wellness, both physical and mental. It takes years for the body to build up cellular defects, which we often dismiss, yet we expect the burden of pain to disappear instantly. Our bodies are made of water as a vital element of life, serving as a transmitter of earth and sky energies and fuelling emotions of joy through the heart. This has been replaced with fake drinks that are just like man-made foods, unknown to cells and body systems made by nature, and spiralling those bodies out of date. While water carries the Earth's memories, nature's system upgrades, and a pearl of immersing wisdom, it also activates the power of knowing the secrets of life and making the right choices.

Human emotions are triggered by the internal environment. Its origin is an intricate balance of elements and cell memory, expanding the senses that have always existed within. Mental and physical health are an inseparable part of it. Humanity certainly had extraordinary achievements over the last century, but it also paralysed our natural abilities.

All inner senses are created by nature and meant to sharpen the ability to go inward, where inner guidance resides. Just like birds, animals, sea life, or land life have inner senses to navigate, know nature's ways, and feel life, so do we. Just like dogs know the owner is coming home way before the car is parked in a driveway, birds navigate their directions, know the wind, and are familiar with all weather and seasons. We, too, have all that coordination built into us by nature within our smart cells.

Our bodies are nature's extension of the earth, like all life forms, and bound by the laws of nature. It isn't something that needs to be proved, it's the way it is! Just like one cannot deny the fact that jumping from a tall building or taking a poison naturally will grant its body death, denying the laws of nature and its direct tie will do too. No one needs years of research and science to prove that it's called common sense. Many have faith in God, Source, or Divine Creation or believe in a higher power or cause incidences, serendipities and so on.

The point is, whatever it is; it exists and there is no denial about that. Just like one cannot deny, it has to breathe to sustain life. Whatever it is to you, it flows through you and every cell of your human body and all living matter, guided and steered by the force of life; where we must go and the choices we must take to evolve as a species. For centuries, we explored and mastered the world outside of us without knowing cellular networks and what is possible inside of us.

While initially, we got the sense of achievements in food innovations and our abilities to store and preserve foods driven by the wish to illuminate hunger, some achievements are proven to be not such a good idea after all.

As human food innovation thrives, nature's innovation declines, causing gap defects and split in systems' original design, each going separate ways, while it is only one design can possibly stay. That design is our origin and the one that will always work in line with the Laws of Nature. It is due to no limitations in food innovations, without realising it, so many cells and cellular networks are as polluted as the planet, air, water, and soil and are in desperate need of cleaning up, interconnecting, remembering, acknowledging their abilities, and retraining to restore and sharpen their senses and abilities because they surely have far greater use and capabilities than we think.

Something that was known as hunger years ago happened to be fasting in the modern world, which brings nothing but benefits to human health. Go figure! I often feel like it is time for all of us to go through everything in our system, laws, productions, education, tech and innovations with a fine pick and review of who it serves, how relevant it is and what it brings to the future that we want to create for us and all species? Because it is all species, co-existence creates an environment: favourable or distractive.

Your own Alchemy of Life GPS, Nature's Essence, which is billions of years of old data residing in your NatalDNA template cell memory, is the connection key to all possible guidance and abilities that lie at your disposal by going inward via pathways residing within your trillion cells and discovering what is possible. Or do nothing and continue existing as half of your potential, searching for what is a natural, high, and indefinite state of bliss and joy while living in an outdated system that the human race has outgrown.

It is said: "One achieved the state of true growth, wisdom and value for the divinity when no longer fascinated with the accumulation of staff, but being in a moment experience each as it comes. Mustering magic and magical moments that come with it more frequently."

My daily wonder sets my soul on fire with excitement and blissful joy, knowing that if we were able to create what we have with all the physical, what is possible to create if we transfer our focus and effort into the nonphysical of living matter and only what is needed as pro-life and as a unified field? Air pollution and Global warming are only the planet's responses to what we do wrong. The moment we focus on inner architectural health, the sooner reflection of that environment will manifest itself on Earth. This is simply because, through our actions, we no longer fund or give out our energy to something that does not serve our health, purpose, or the health of the Planet. What is no longer in demand, just like old music tapes, will spiral itself out of our lives, taking all that is wasteful, pathogenic, and not pro-life with it. Although it is always free will" for all, many pieces of the puzzle contribute, and one must take action to obtain understanding or, at the very least, ignite the possibility of such potential before fully fledged power is achieved. After all, no one can say it isn't possible because, two hundred years ago, walking into space was also viewed as impossible.

The more I see Nature as an Architect of Life, the more I see our failure as a species that unknowingly wandered off and somehow lost its way in a system that has no future on this planet, like unknowingly playing for the wrong team or running someone else's marathon. We are at the peak of technological developments, yet we have stopped evolving as a species and as living things. From an evolutionary standpoint, we have accelerated decline rather than growth.

The difference between a healthy cell and a cancer cell is that a healthy cell will only take what is needed for that day, take its mastery, and be given time to accommodate existing and create a new life with it, making every element count and making the most of it, while a cancer cell will consume anything and everything with a never-ending appetite for consuming to consume—a pathogen in nature. Even fat cells are only stored because the cells do not recognise them as beneficial building blocks or know what to do with them, tossing them aside as storage until better days. Like soil and plastic. Where the soil doesn't have tools to break down the plastic, because it isn't natural, nor was it grown by nature. Fat cells and everything in them are stored because they have no elements to convert them, and although they cost energy and are hard to keep, they're almost impossible to dump without need. It robs energy that is produced by healthy cells, depriving the quality of life the cellular team of that body could have otherwise. The trigger is an environment that is separate from the original by design, pretending to be nature while it's not, like a wolf in sheep's skin.

Things we developed like artificial additives and flavourings trigger changes in behaviour, causing us to eat more, consume more, and care for nature given body less than all the stuff we possess, without consideration for the planet, funding goods and services that create waste and contribute to more waste at a faster pace than the Planet can break it down and keep up with. Without understanding that even biodegradable in large quantities is still too much, our pathogenic behaviour is too much.

Just like the food, supplements, and vitamins that may be good for us in the wrong quantities, they are still too much if the body must work overtime to break them down, causing accelerated self-inflicted decline over time. And, just like the human body, which increases its temperature to an extreme in the hope of destroying the highest level of pathogens that are putting the body in danger, the Planet does the same to reduce pathogens and pathogenic behaviour—which is what we know now as Global warming, but it doesn't end there. What comes after the human body reaches high temperature to suppress pathogen spread? In the healthy body, it follows severe sweating to achieve cooling. In Nature earthquakes and volcanoes represent such function, with one variable, that is how much ash clouds are released to tackle cooling down, which in large quantities or in small catastrophic global events can very well take us to the ice age.

Nature is a living consciousness that expresses itself through the living matter. As I have mentioned, our bodies are nature, an extension of the Earth and Nature's all living matter. In many ways, operates just like us, knowing what we know because it upgrades through us. Just like human consciousness reads local and global data of its own environment and implements adjustments accordingly. And just like you would know, if you have a temperature, are feeling unwell, have a headache, aches, pains and so on, Nature does too.

The same goes for knowing humans like her children and treating them with the most tolerance until they go too far. Naturally, more advanced, and more intelligent than we are because she is very attuned to all living matter, including human thoughts, the energy of the heart, true feelings, values, and behaviour. Patiently navigating through its laws to sustain balance for and within all life forms.

Nature's current state of assessment and observation focuses on hope, allowing us to run on borrowed time wondering if we are willing and capable to pivot, and not for herself but for us. After all, it needs only a few species to restart again should it come to do what it must. Even if Nature does nothing our current state of genetic defects, illnesses, infertility and so on will become more prominent with a fast decline in life span. This is because the human genetic template was not designed to work on manufactured and modified foods so-called "garbage "that Nature does not understand. Nature's preference would be not to waste the benevolent achievement of humankind and a global rise in human consciousness as there were good achievements among bad ones. Some achievements are remarkable, but not fit to work with the current system based on human greed.

Even if Nature knows what we do wrong, it will pretend it knows nothing and simply wait, because many of us already walking the path of self-created slaughterhouses. Living in a world where we have and can create anything, it is no longer about what we can do or have. However, it is about knowing what we can do or have, but choosing not to because the other life would be effected by us making that choice. Such questions must be asked at every choice we make daily and at any step of any product, service development or innovation. How would it affect nature's resources, environment, soil, an animal, insect or a stranger, the future of accumulated waste, and every particle of life with every choice we make? Chemo treatment we have developed to treat pathogenic behaviours in cancer cells has already made its way into our lives driven by human ignorance, a sense of invincibility, greed for profit or lack of understanding.

Where many willingly consume chemicals, knowingly or unknowingly, so they get hooked on them while bathing in radiation with a "know-it-all" attitude and are so unwilling to give up. Quietly and surely bringing the body expiry date as close as possible. On the contrary, the early expiry date of nature's given body at a current state is more useful to nature in the form of compost to replace overconsumption of what was taken with less ill and incapable humans to feed. Nature has no interest in keeping us with our current system as it is - it does not serve her to do so, and it makes her ill.

Remember, Nature's most powerful Law of cause and effect is always in play and guided by the Law of Compensation. Nature built humanity as superior among all species for one purpose only: to guard, evolve, create, and innovate to help Nature preserve resources, cultivate and reciprocate life force energy, master the power of Natural Laws and support all life forms, not to destroy them.

16 CHAPTER

THE POWER OF NATURE'S EVERLASTING LIVING

MASTERY

The value of the human body, the power of human senses, abilities to feel life, life force energy and connection among all life is not only priceless but truly magical. After all, it is the very reason we, as souls, come to this life, agreeing to almost any conditions that come with it. Even knowing it will be challenging, we may not like everything about it and that it will end as quickly as a destination trip because it is bound by the Natural laws of the living matter, our souls still queuing to come back. Why is it so we will expand more in-depth in the next book, where we will be diving into the alchemy of souls and energy, frequency, tapestry, formation, density, purpose, power, and expansion?

Everything eventually will make sense, why life is the way it is, why it rarely repeats the same, why we are the way we are, the power we hold within us, and why our body's cellular team is of the most importance and relationship with must be mastered to create the world and the life we all want. Fundamentals and understanding why and how the soul and the body of the living matter are interconnected is an inseparable part of the evolutionary growth of any life form, which comes not without a choice. Just as cells build and form perfection and healthy functions of the body, it is we who do the same for the healthy function of this world. I want to share highly unusual experiences that may help to grasp an understanding of some topics covered in this book.

When I was sixteen years old, one Sunday morning with nothing unusual, I was having a cup of tea with a slice of lemon in it when an autoimmune allergic reaction of anaphylaxis struck. Bear in mind, I've been having tea and lemons as long as can remember and never had a particular problem, except a few times in the past when I woke up with a swollen cheek without apparent reason. My body started to swell, going into shock and closing my airways, and within half an hour to an hour, I was not breathing. From what I can remember and from that point onwards, I was observing everything from outside the body. I was watching paramedics, my mother, and everything that was happening to save my life. Shock must have thrown my soul out of my body. It felt like it was needed to prevent damage. I came recently to understand the why behind it. The experience of clinical death showcased an incredible perspective in my life, for which I will be forever grateful.

It wasn't the only time it had happened since, but this one engraved in my memory forever. When it happened, it wasn't the death that was fearful or that scared me because I felt nothing, which would be the case under the circumstances of the shock. It was neutral, peaceful but cold like indifference and inability to feel anything.

The best analogy I could come up with is if I were watching an emotionally epic film or wanting to smell flowers, taking a big breath, or attempting to feel the sun or the wind on my skin, but couldn't, feeling absolutely nothing, yet remembering how it all feels. All these memories and cravings to feel when I eventually came back into the body changed my outlook on life forever.

Nature of the living matter, especially the human body exists, just so one can experience all emotions and beauty of life because such is impossible when you are just a soul. Which is as powerful as an engine yet not as powerful as when it is part of a starship. Such a gift is given to souls in the form of the body by Nature and is only achievable via Natural Laws with its architectural and alchemical essence at the fundamentals of the living matter. It is the unspoken relationship between both, which allows us to experience life every time we are desperately keen to come back into it. Everything in the universe is in constant change between form and formless, just as death and rebirth or dark and light. The value comes through understanding which one we feed.

When I was out of my body, my soul vision was incredible, I could see at 360 degrees and beyond, moving instantly just with a thought, but I couldn't feel the temperature or any of my senses. It was the state of the observer and felt as if it was neither past nor future, but only now. It was hard to hold any thoughts. I felt as light as a feather and as if it be easy just to fly away. As I was about to leave my life, I was distracted by an intense sound of drums mixed with an aggressive, unbearably loud sound of a powerful but somehow distorted guitar.

The Sound felt like a magnet dragging me back and stopping me from leaving. It was so loud that it was impossible to ignore, the sound and frequency of it pulled me back into my body. When I finally came around recalling everything, I realized though it wasn't my time, I must not take life for granted, and learn all I can, especially to enjoy every given moment. Life isn't about accumulating stuff, it's about the experience that adds density to the soul's growth, giving rise to the emotional intelligence we know as wisdom. When I came around, the hospital was the same as I'd seen it while witnessing it all. From that moment forward I've looked at my life as if it was a raindrop in the ocean of life inseparable from air, water, earth, plant or a dragonfly and every little step to preserve life is a matter contributing its power to the collective.

We are only as important as reciprocation from others in the form of attention or admiration making us feel such a way. Simply because when we do so we share our life force energy with that person making them light up with the accumulated power we give. Nature's Divine living matter does the same for all of us.

Being an empath, the cold-like indifference of the void made an impression on me, leaving me with a slight chill every time reminding the importance of every single breath and moment being directed well when one shares such precious power. And I rather share it with nature and those who care, and support life as opposed to those who do not. I started to value everything about life, all moments, and all life forms, Nature, and our planet, learning all I could about nature's ways, medicines herbal and modern, ancient teachings and scripts, holistic methods, Natural Laws and order, the nature of human body and living matter. Knowing I may have another life, I made a conscious decision to cherish this one and the body I was given with every breath and share it with others or perhaps help as many as possible to understand the whys of life, the true power of the human body and what makes it all so wonderful and priceless. No one was telling me what to learn I was simply allowing to be guided from the heart. We stop growing and evolving as a species, when we stop caring and step into the "rat race" which is out of sync with our natural pace and the source of origin. Inevitably, we lose interest in learning; we stop loving our work or what we do, and we get exhausted with life assuming we know everything, creating a mental block to possibilities. Without realising that system that exhaust us been created and encouraged by our own participation. Although change is an inevitable trigger for growth, it is the Natural pace and environment that are the key to inspiration.

It was interesting to learn, after twenty years of avoiding all citrus fruit and citric acid and after recent years reverting to full commitment to my NatalDNA diet, lemon is back into my life as an inseparable part of my daily elements. It is also inspiring to observe that many holistic treatments, natural solutions and old ways of finding natural pace, fasting, breathing, meditations, energy work, hyperbaric chambers, ice baths, sound therapies, herbs, and remedies that I've grown up with are slowly coming back to our world and the world of wellness. It is we who create demand and it is we who make shops, farmers, businesses, and manufacturers change their ways, quality and packaging.

As I said earlier, of course, we can carry on as we are, because we can, but it evolution of the mind among all species allows us to recognise our power as one when we tackle daily choices, minuscular or grand. In nature, views and actions change as observation changes and as we are nature, we should attempt too. After all, no harm in trying.

For Natural Laws to gift life filled with reciprocity of joy wonder and magic, one must simply take time to notice and learn its ways, so one can achieve the highest state of high making the impossible possible while granting us with its majestic presence.

The dance among all life elements is at the core of essence and governed by the Laws of Nature. The overflowing state of 'I want' over the state of 'We have' cannot ever be discarded without suffering its effects. Just like one cannot avoid flooding after wanting too much rain or burning after using up reserves of elements that offer protection and buffer the heat from the sun. Where there is pull, there is a push in the intricate design of life.

Nature will always compensate to sustain the Law of Balance. Which way it will compensate, your health or the health of your children or other term oils in your life will depend on the choices you make daily and the way you view all life forms from the perspective of being as grand as Creator or as minuscular as a cell or fungi in the human body. Considering everything we discussed in this book. Nature's creation, gratitude and compassion towards all life forms include every single cell in the body that you borrowed from nature to experience this life as a soul, acknowledge and respect Natural Laws and view everything from the perspective as if you were a single cell in your body among trillion life forms influencing the outcome of internal and external environment it creates for itself and others, which cellular behaviour would you rather adopt? The pathogenic or life-thriving force of life with limitless abilities but I assure you the law of balance will always compensate above all.

Regret is the last-time, last-minute effect of a cytokine storm of an expiration date from a prior given time, which at that stage is no longer reversible. Understanding regret and its superpower that has been given to humans to master, so one can never live to experience, is a true gift bestowed by Nature.

Ancient knowledge, passed down through the ages via various pathways, holds the wisdom of understanding the Natural and Universal Laws that serve as the foundation of Nature's Essence, Power of Health Energy, Alchemy, and your role in this world. Where your presence is as minuscular as a form of cell in your body or as a raindrop but as important as a bug in nature and the power it forms collectively to create oceans of cellular intelligence, creating life force energy that moves it all.

It reveals Nature's role as the genius Architect of life, embodying an essence that transcends time and space. Embracing this wisdom allows individuals to develop a remarkable ability to perceive, to see beyond the surface, and to evolve their outlook from different vantage points and perspectives. It grants them the gift of discerning what is right or wrong in alignment with the intricate workings of the natural world.

For instance, organic foods contain nine times more nutrients than nonorganic ones. When the body's system is efficient at self-regulating processes to break down, convert, and absorb foods that nature intended, the need for food and waste naturally diminishes. Herbs have been provided by nature to aid any necessary modifications because they are recognised by our bodies, which are a part of nature. Having one or two meals with higher quality and compatibility foods in your diet, the easier for the body to navigate. Means it has to work less and it recognises the foods of the land allowing cells to thrive with life. All this precision in food choices serves to nourish us and provide the energy needed by the trillions of cells in our bodies. If cells don't produce a sufficient amount of energy, their function will be limited – this is where fatigue stems from and the illness forms.

All living matter is fused with life force energy and the majority is produced within each cell. What is the opposite? The answer is nothing, or what ancient knew as Ether, shadow space, the opposite of life and what moves life. One does not exist without the other. When living matter is no more, it is released into particles to be recycled to fuel new living matter. Nature and its Laws are architected to keep order on this Planet to sustain life and be a home for living matter. And for us, come to borrow such living intelligence and experience life in the bodies we are given. That's what makes the difference between a living planet and a dead one. Just like a borrowed car and you as a driver, once the use of experience is over, your soul will return to space the ocean of energy consciousness until the next time. Like a drop into the ocean. Remaining all the memories yet unable to feel them. Not feeling pain, but not feeling love or joy either.

Hypothetically speaking, if a group of 'smart foods' compatible with one's NatalDNA template could produce 100 units of energy per cell instead of just twenty units, the need to dip into backup resources and the rate of wear and tear would decrease. This would improve our repair capabilities and save time, all while preventing waste and the unnecessary pollution that comes from manufacturing food that is useless for such purposes. This would allow humans to explore other abilities that nature may have planned for us if we dare to take the chance to evolve in that direction.

There is something else. Before, I adhered to the NatalDNA diet. I was always cold. Which was thought of as poor circulation with no solution and absolute norm. When my diet was not in place, my electric bills were sky-high as I struggled to keep warm. I used to think it was due to the poor circulation. Well, now I know it was the diet that was not suitable for my body, increasing my insulin resistance, and making my feet or hands often feel numb. It was horrible to feel the cold to the bone and have to jump into a red hot bath at any opportunity, just to warm up. Some of us are prone to feeling cold by nature some of us hot, nevertheless, no one should suffer to experience any extreme side of it. It is unhealthy and it is fuelled by food and deficiencies which can be corrected with persistence, patience and time.

Now I never have those extreme symptoms, nor do I feel the cold or need to have my heating in the house sky high. Naturally, such a strategy will help easily implement a reduction in the need and usage of gas/electricity needs to those who are interested in achieving that. Inherently, once any condition develops it cannot possibly, magically disappear. We cannot force the pace of nature and should know it will take time to repair. One should be honest with oneself that if it takes 20-30 years to create damage for the condition to develop, it may take at least seven years to repair it with continued minor improvements each year.

Just as we cannot plant a seed and receive a full-fledged plant the next day or even in a few days. Why seven years? In nature is the full body replacement growth cycle from blood cells to the bone. Naturally, just as we can't skip seasons or years, or architectural fundamentals in building anything, the body cannot skip a natural repair cycle. Making preventions is the best solution and living Mastery of all.

By delving into this ancient knowledge, we unlock the secrets of Nature's genius. We learn that the laws governing the universe are not arbitrary or capricious, but rather systematic and purposeful. They offer a framework through which we can navigate the complexities of life and understand the profound interconnections that bind all living beings.

With this understanding, our perception expands, enabling us to transcend the limitations of a singular point of view. We gain the ability to see the interconnectedness of all things and to recognise the delicate balance that exists within the web of life. From this elevated perspective, we can discern the inherent harmony and order in the natural world.

As we align ourselves with the Natural and Universal Laws, we tap into the vast reservoir of wisdom that resides within us. We stop funding our own illnesses. We stop the overuse and over-exhaustion of natural resources and access an intuitive knowing, an inner compass that guides us towards what is in line with these laws. This deep connection to the essence of Nature's Alchemy empowers us to make choices that are aligned with the greater good, choices that nurture life, promote balance, and honour the interconnectedness of all beings.

Through the lens of this ancient wisdom, we gain a profound appreciation for the intricate beauty and intelligence that permeates every facet of existence. We recognise the inherent brilliance of Nature's design, the wisdom that is woven into the very fabric of life itself. By embracing this wisdom, we unlock our own potential to become co-creators, harmonising with the natural order and contributing to the evolution of consciousness.

In a world that often seems chaotic and disconnected, the wisdom of Understanding Natural and Universal Laws offers us a guiding light. It provides us with a compass to navigate the complexities of our modern existence, to make choices that honour the interplay of all life forms, and our unique unexplored talents and to cultivate a deep reverence for the genius of Nature. Let us embrace this ancient knowledge, integrate it into our lives, and embark on a journey of profound transformation and harmony with the wisdom that has been entrusted to us through the ages. After all, it is single water drops that form oceans of unity and hold the power of life force energy among all life forms the most, but it is Natural Laws that guide the way.

Given everything in this book, Nature gives us a chance to pivot. What we do with her guidance and the daily choices we make is up to us. It is freedom of choice, always. Remember, you are nature and nothing in nature is done by force. The Power of Nature's Everlasting Living Mastery

A Dose of Common Sense

What often goes unaddressed—something that many healthcare professionals may not fully comprehend—is that the incidence of most health issues is on the rise precisely as we invest trillions in "researching" and "treating" them. While funds can be used for creating healthier communities using preventative holistic methods and reducing the cost of living for all. In other words, if the country's population pivots and reduces its illnesses, there is no need to fund research and treatment of such illnesses.

Amid these unprecedented patterns affecting our mental and physical well-being throughout our lives, all rooted in metabolic dysfunction, we are urged to "trust the science." Clearly, this is illogical. Over the last sixty years, we've been discouraged from asking questions just as rates of chronic diseases and mental health have sky-rocketed.

Naturally, the reality is that we should heed the advice of the medical system in cases of acute issues such as life-threatening infections or fractures. However, when it comes to the persistent conditions that disrupt our lives, scepticism is warranted for nearly every institution offering guidance. The solutions are much simpler and within our control. Having an understanding of the Natural Laws, precision, "cause and effect", taking responsibility and control, prioritising your food choices for the health of your nature-given body and guiding your family into the future is most crucial.

Just like plants draw nutrients from bioavailability in the soil, we do too, from bioavailability in the blood after the body makes it. More so, each organ cell would draw its unique formula of required elements, which don't have to be bio-available from a mass variety of food resources. The larger and more diverse daily food variety is, the harder the body has to work, adding more confusion to the body system narratives.

For instance, when one uses one type of quality protein that goes with an allocated range of vegetables, it helps to utilise daily building blocks effectively and without waste, condensing toxicity to as little as 5-10% as opposed to 50 -

70% when a wider variety of food in the one's diet. Naturally, this allows the cellular team of that body to work without extra strain and over-exhaustion, allowing room for energy cultivation abilities and other functions.

The analogy is relevant to how much chemically dependent or waste-generating manufacturing we seem to support by funding indirect toxicity via creating demand. The higher the number of such manufacturers, the higher toxicity and waste would be afloat in the pool of economy and life, adding its share of toxicity. The lower the demand driven by fewer narratives, the lesser the toxicity levels would be. Such understanding allows us to select pro-life narratives for balanced and pro-environment industries that we choose to fund in the future.

For the purpose of an example, would you choose a destination trip or pack of crisps daily if you knew emissions from each nearby were the same? When one pack of crisps, sweets, nuts and other packaged foods that seems innocent contributing to toxicity every day. Let's explore how damaging this emission is on a grander scale overall. Numbers are approximate, and only to exhibit indirect damage we fund contributing to human health and the environment that serves as a cause of failing health in the years to come.

Weekly habits of bagged snacks in the UK currently stand at around 58%, used twice per week, which may sound not significant at first, but it is around 30 million. Which on its own perhaps not a big deal; however, if we are to expand and think about how much waste and packaging it creates during manufacturing, including CO_2 via deliveries of raw and ready material to and from manufacturing, workers travelling to and from manufacturing, deliveries to supply hub or shops to and from and so on.

Worth mentioning coffee is harmful for some but not others so establishing individual NatalDNA needs would help with health implications caused by waste. Also, it shouldn't come in instant form or capsules of any shape due to the waste it creates even when recycled. Organic coffee beans are best in their natural form and should be grind preferably prior to the time of use. Such helps to reduce the need for other forms, and additives, knowing exactly what's in it. All of the above soon adds up, including its actual effect on individual health, providing us with a bigger picture.

Having leaders with life experience in different walks of life brings a profound level of comprehension and ability to build local and global health for all. How does one find the solution for the circumstances and situations that one has never been in or never experienced?

A system that allows opportunities and flexibility to experience different industries and all walks of life removes rigid mentality and develops the flexibility of the mind by providing an understanding of cause and effect, without imposing on the freedom of choice.

Precision matter. Some countries support the health and wellness of their citizens by implementing a profitability cap (profit limitation) without taxation on life essentials that support the popularity of healthy habits to help the growth of a healthy population. Essentials, such as organic foods especially local and country own, holistic treatments, the field of wellness and education on holistic ways of living, community wellness, green building, behaviour towards energy saving and healthy living incentives, reduced working hours and environment with the room for required diet, home cooking and green transport. Why this is important?

When such a structure is in place, it supports natural health, NatalDNA diets, fitness, and wellness. Organisations offer six monthly holistic health or spa retreats as opposed to health insurance or health insurance that offers such a service. It means less need for funding researches for treatments and illnesses we create, creating reduced need for hospital beds, care homes, medications and else. Or instead of research that is directed to research and prove the obvious, instead of spending funds with better means for children and adults or else. Such pivot allows funds redistribution where it needs to be rebuilt and not where it is meant to be patched up, directed towards healthier ways of living.

Mindful funding choices and allocations for innovations. Nature's point of obvious is for the creator not to be engrossed in the genius of its own greatness by creating for the sake of creating with one-sided interest.

For instance, instead of funding holistic health and organic local farming as preventative component and help in rasing healthy humans. We often fund tech to help the health system cope with volumes, without realising, that many who are in ill health at the other end (patients) will have difficulty navigating through mobility of their own body or thought pattern, no matter how advanced tech is. When one struggles to connect thoughts while in pain. How is the tech system that requires logging in and going through many different security protocols would help the person who is ill? Human beings who are elderly, ill or in pain and have no physical or mental capacity to be conscious of their behaviours at a time of need due to pain, as opposed to one that is healthy and capable of grasping tech. Why is it not obvious that such tech is useless to support wellness and care for those in need?

Or when males assess female health and vice versa while neither lived in that gender body to have the actual understanding of how it works, making it based on pure assumption.

Public transport, underground and railway should be not for profit. With profitability cup applicable in other industries, especially those contributing to wellness of the society, reduced pollution and clean environment.

The importance of a conscious living, conscious system, conscious choices or practice either locally or globally is as important as the air we breathe, it becomes us.

The future of conscious schools. In the era of technologies and society that care for wellness, successful children with high grades can be rewarded with permission to take up to four weeks of holidays throughout the year. Such a challenge is rewarding and encouraging for parents as well as children, especially if they are learning about different cultures and developing the flexibility of the mind. When the system is considerate, people are happier, creating greener ways of living and levelling economy. Where possible, students can have a choice and freedom to develop discipline and have freedom to take lessons remotely or in class. With options for remote studies or in person. There are many ways to explore this dose of common sense.

Where possible support local organic farms with nature-guided ethics. Purchase with as little packaging as possible to reduce the need for recycling. Buying with needs rather than wants and buying quality that lasts.

The neutrality of NatalDNA diet is truly remarkable. Due to my understanding of Natural Laws, mechanics of the nature of the NatalDNA diet and how nature's allocation works, I restrain making recommendations without knowing the personal history of each unique body and its origin. Precision in supplement's matters too. However, here are simple guidelines that anyone can attempt to try:

1. Aim for organic, simplicity and purity by removing anything processed, manufactured or that has more than a few ingredients in it. Avoid foods with any ingredients or E-numbers that you do not understand or, don't know if it's compatible in its form. Anything long-lasting would have preservatives in it or grow bacteria.

2. Any form of indigestion indicates either incompatibility of foods, drinks, low stomach acid, high stomach PH, toxic metals or lacking foods that serve as converters.

3. Foods to remove for 3-6 months to achieve clarity on your body NatalDNA compatibility with those foods. Which, when tried one by one at a later date offer sharper body signaling communication. I recommend removing all deadly nightshade vegetables (tomatoes potatoes eggplant peppers including chilies), all wheat, all soya, rice, all grains and legumes, all dairy, broccoli, all GMO foods where possible, all sugar including sugars in foods or natural sugars such as honey or fruit (berries are ok), alcohol, juices, and drinks with any form of additives or flavourings even natural and unfiltered water. When reintroduced after 3-6 months, if there are no

underlying issues the body will guide you by telling you what works best as the best form of fuel.

4. Remove face and body creams and replace with natural substitute. If you can eat ingredients, it means it's natural. Watch out for fluoride-free natural toothpaste, choose holistic toxic-free teeth fillings, chemicals in cosmetics and everything that goes on the skin. Look for natural deodorant and body fragrances. Think of it as if you were putting it onto the wound or the soil where you will be growing something healthy. Ask questions.

5. Learn and practice fasting or use guided fasting, if unsure. Gradually building into 18 hours out of 24 hours. Day at a time. Ensure the sufficiency of electrolytes and quality of compatible foods. Mindful of herbal teas, spices, seeds, nuts, and complex supplements as they may not be compatible with your NatalDNA template. With any history of ill health must consult your medical practitioner.

6. Keep it natural and simple. Remember, if one slice of lemon almost ended my life by changing the behaviour of my cellular body environment; what else is possible, when complex compositions are in place? Think of everything from the perspective of cells. Is the product produced by nature, created by you, and fresh?

7. Practice breathing to expand lung capacity. Oxygen has remarkable benefits and require regulars practice and mastery.

8. Spend time in Nature, seeing her beauty and life she creates. Adopt a positive attitude and daily gratitude by focusing on what you have and what an amazing, special, and unique creation you are! Be the love you want to see in the world.

When progress is defined by the growing numbers of poverty, unhealthy beings, disability and demand for care homes and hospital beds, one must ask questions that lead to answers at a source. The message emphasises the power that lies within each individual as the origin and pathway displayed in the principle of infinite global health designed by nature at the source for all living matter, which is simply given to all, bound by, and defined in Nature's Architectural Laws.

Using Nature's perspectives of our origin as intelligent species, we can focus on importance to support preordained wellness within ourselves and all life forms by educating ourselves and making right choices; by establishing what drives degradation of our unique abilities as species and use preventative

methods to address it. Regardless of the origin of such being introduced in our lives as a consumer. We must choose the best solutions, whether it be foods, lifestyle, habits, mindfulness, goods and poor quality goods, technologies that stop us from using our brain or our thinking from a peripheral perspective and remove human interactions or valuable jobs. Because only an approach of such nature will stop humanity from developing backwords as species.

Nature's rewards. When one follow the body manual, its unique NatalDNA Diet allocation and adhere to natural Laws, higher faculties to life's wisdom and collective consciousness become open to facilitate true purpose and unique physical and spiritual evolutionary growth opening pathways to higher intelligence. Transcriptions from collective consciousness 'the voice' become open and available for unique guidance after energy pathways are cleared, energy cultivation in place and expansion through reciprocation is achieved.

ABOUT THE AUTHOR

I am profoundly, irrevocably and unconditionally in love with the Nature of our beautiful blue Planet Earth. I've lived and called many lands home and experienced countless and diverse ways of life and for all that was offered, either joyful, miraculous or challenging, I am forever grateful. The Power of Universal Energy resides in Cellular Intelligence within us, expresses itself as us and most powerful in Unity as One.

"Love, Compassion and Kindness are action words with no boundaries. All limitations are personal barriers which begin in the mind but dispense in the root of Essence. Retrieve your Nature Given Gifts and Building blocks of Life – The place where Health, Joy and Love reside."

Natalie Manroe

Printed in Great Britain
by Amazon